RISK CRITICISM

Risk Criticism

**PRECAUTIONARY READING
IN AN AGE OF
ENVIRONMENTAL UNCERTAINTY**

Molly Wallace

UNIVERSITY OF MICHIGAN PRESS
Ann Arbor

Copyright © 2016 by Molly Wallace
All rights reserved

This book may not be reproduced, in whole or in part, including illustrations, in any form (beyond that copying permitted by Sections 107 and 108 of the U.S. Copyright Law and except by reviewers for the public press), without written permission from the publisher.

Published in the United States of America by the
University of Michigan Press
Manufactured in the United States of America
⊚ Printed on acid-free paper

2019   2018   2017   2016   4   3   2   1

A CIP catalog record for this book is available from the British Library.

Library of Congress Cataloging-in-Publication Data

Names: Wallace, Molly, author.
Title: Risk criticism : precautionary reading in an age of environmental uncertainty / Molly Wallace.
Description: Ann Arbor : University of Michigan Press, 2016. | Includes bibliographical references and index.
Identifiers: LCCN 2015038637 | ISBN 9780472073023 (hardback) | ISBN 9780472053025 (paperback) | ISBN 9780472121694 (ebook)
Subjects: LCSH: Ecocriticism. | Criticism. | Risk in literature. | Environmental risk assessment. | Risk-taking (Psychology)
Classification: LCC PN98.E36 W35   2016 | DDC 809/.93355—dc23
LC record available at http://lccn.loc.gov/2015038637

For my family

Acknowledgments

This book has been a number of years in the writing, and I am deeply grateful for the support and encouragement that I have received along the way. To my colleagues at Queen's University, I owe a debt of gratitude for your support and inspiration. Thanks particularly to Mick Smith in the School of Environmental Studies and Asha Varadharajan in the Department of English, both of whom let me test out some ideas on their unwitting students. And thanks too to the students in my own courses on "nuclear culture" and "risk" for thoughtful engagement and challenges. In the wider world, thanks are due to Rob Nixon, for his encouragement and inspiration, and to Begoña Simal-González, whose invitation to deliver a lecture in Spain seemed too good to be true, and who, in addition to friendship, offered the opportunity to share portions of what is now chapter 3. I am also deeply grateful to Catriona Sandilands for her immense intellectual and personal generosity, including the invitation to deliver a closing keynote address at the Green Words / Green Worlds conference in Toronto, which gave me the opportunity to craft an early version of chapter 5. A version of that keynote (and now a proto-version of the final chapter of this book) is forthcoming, as "Averting Environmental Catastrophe in Time: Staging the Uranium Cycle," in a volume that Cate and Amanda Di Battista are coediting for Wilfrid Laurier Press; my thanks to the Press for permission to include a version of it here. Other portions of the book have also appeared elsewhere. A variant of the introduction appeared as "Will the Apocalypse Have Been Now? Literary Criticism in an Age of Global Risk" in *Criticism, Crisis, and Contemporary Narrative: Textual Horizons in an Age of Global Risk*, edited by Paul Crosthwaite (Routledge, 2011). My thanks to Paul for organizing the marvelous conference on risk for which the paper was originally written. An earlier version of chapter 3 appeared as "Discomfort Food: Analogy, Biotechnology, and Risk in Ruth Ozeki's *All Over Creation*" in *Arizona Quarterly* 67.4 (Winter 2011).

My thanks as well to those who have provided permission for the images in this book, the *Bulletin of the Atomic Scientists*; the Argonne National

Library; Heal the Bay; Boom Entertainment (for the image from *I'm Not a Plastic Bag* © Rachel Allison. All rights reserved. Used with permission); John Hendrix; the Maruki Gallery for the Hiroshima Panels; and John Junkerman (and the Maruki Film Project). I am grateful also to Adam Dickinson and House of Anansi Press and Ronald Wallace and the University of Pittsburgh Press for granting permission to reprint poems. *Risk Criticism* was supported with a Standard Research Grant from the Social Sciences and Humanities Research Council of Canada (SSHRC). My thanks to my Research Assistants who contributed immensely to the project: Maryanne Laurico, Laura McGavin, Aaron Mauro, David Carruthers, and Shadi Ghazimoradi. The staff at the University of Michigan Press has been unfailingly professional and tremendously supportive. Thanks especially to Aaron McCollough for his support of the book, to the two anonymous reviewers for their generous engagement with the manuscript, and to those on the Press Committee, who provided wise council for final revisions.

This project would not have been possible without the guidance and example of my parents, whose commitment to literature and justice is a constant inspiration, and whose loving support has sustained me. I owe many thanks to Todd, builder of native gardens, birdhouses, and rocket-mass heaters, who has championed me unflaggingly, and to Lucy, lover of worms, birds, cats, humans, and all other planetary life, who is teaching me hope. As I was readying the manuscript for submission, I received word that Ulrich Beck, the source of so much inspiration for my work, had died. I would like also, then, to acknowledge the deep debt I owe to him, and offer thanks for his life and work.

This research was supported by the Social Sciences and Humanities Research Council of Canada.

Contents

INTRODUCTION
Will the Apocalypse Have Been Now?
Literary Criticism in an Age of Global Risk ... 1

ONE
The Second Nuclear Age and Its Wagers: Archival Reflexions ... 28

TWO
We All Live in Bhopal? Staging Global Risk ... 64

THREE
Discomfort Food: Analogy and Biotechnology ... 93

FOUR
Letting Plastic Have Its Say; or, Plastic's Tell ... 123

FIVE
The Port Radium Paradigm; or, Fukushima in a
Changing Climate ... 154

AFTERWORD
Writing "The Bomb": Inheritances in the Anthropocene ... 192

Notes ... 209

Bibliography ... 239

Index ... 255

INTRODUCTION. **Will the Apocalypse Have Been Now? Literary Criticism in an Age of Global Risk**

> *Given its impalpability, its lubricity, can this protracted apocalypse be grasped, or only sensed faintly as we slip listlessly through it?*
> —Andrew McMurray, "The Slow Apocalypse"[1]

> *The future inhabits the present, yet it also has not yet come— rather like the way toxics inhabit the bodies of those exposed, setting up the future, but not yet manifest as disease, or even as an origin from which a specific and known disease will come.*
> —Kim Fortun, Advocacy after Bhopal[2]

CRITICAL PRACTICE IN THE SECOND NUCLEAR AGE

From the start of what, in retrospect, may have been the first nuclear age, perhaps no image has so captured the sense of looming risk that nuclear weapons pose as the *Bulletin of the Atomic Scientists*'s "Doomsday Clock," an icon that has graced the cover of that publication since 1947. From its perilously close two minutes to midnight following the detonation of the first thermonuclear bombs, first by the United States and then by the Soviets, in 1953 to its position at a relatively comfortable seventeen minutes to midnight in 1991, the Clock has stood as a barometer of the world's proximity to its end. With the end of the Cold War, this icon might seem to have joined duck-and-cover drills and fallout shelters as an archaic relic of the atomic age; nevertheless, it has continued to mark the times—and has marched fairly steadily toward midnight, from fourteen minutes in 1995, to nine in 1998, to seven in 2002, each tick reminding us that, though the cultural obsession with the nuclear may have waned, we continue to live under the shadow of the atomic bomb.

But even as it represents the continuity of risk, the Clock has also changed with the times. Indeed, when it appeared on the January–February 2007 issue of the publication—reset to five minutes to midnight—its symbolic valence had subtly changed. Still measuring nuclear threats—the United States' then-interest in usable nukes, the spread of weapons to North Korea and potentially Iran, and the resurgence of investment in nuclear power—the Clock had also begun to register other risks that the *Bulletin* felt had graduated to the scale of the nuclear, including particularly climate change, but as the *Bulletin*'s scientific panel of sponsors added, also biotechnology and nanotechnology, an epochal shift that the *Bulletin* suggested constituted a "second nuclear age."[3] As Sir Martin Rees, president of the Royal Society and a *Bulletin* sponsor, put it: "Nuclear weapons still pose the most catastrophic and immediate threat to humanity, but climate change and emerging technologies in the life sciences also have the potential to end civilization as we know it."[4] In the "second nuclear age," then, the term "nuclear" appears to operate as a synecdoche for global environmental risk more generally, what German sociologist Ulrich Beck has called "world risk society."

Periodizing the contemporary is always a tricky combination of divining and conjuring, but whether or not recent events warrant its inauguration, the *Bulletin*'s "second nuclear age" at least offers an occasion for reflection on how we understand contemporary risk. Ticking back and forth between two and seventeen minutes to midnight over the last nearly seven decades, the Clock provides an odd synchronicity, such that, for example, five minutes to midnight put 2007 roughly where the Clock stood in the mid-1980s (between 1984's three minutes to midnight and 1988's six), a coincidence that offers a countertemporality to the successive logic that often characterizes narratives, whether of critical practices or history. How, the Clock invites us to ask, did risk figure in earlier moments of similar proximity to midnight? What might the cultural impact of those risks and the attempts to work through their troubling and potentially promising implications offer us, critics grappling with risk today?

Risk Criticism attempts to puzzle through these questions. Taking inspiration from the *Bulletin* and its suggestive synecdoche, I return to the nuclear, both to emphasize the hazards of its persistence and to illuminate the additional risks that might form the "whole" for which the nuclear part now stands. *Risk Criticism* moves thematically through a series of hazards, potential and realized, including toxic chemicals, biotechnology, plastics, and climate change, in each case mobilizing specific historical examples (from nuclear testing on the Bikini atoll to the Union Carbide disaster in Bhopal to the Pacific garbage patches to the Fukushima accident; from uranium mining to genetically modified organisms [GMOs]) and examining cultural production that grapples with the larger implications of these seemingly disparate events. The *Bulletin*'s second nuclear age thus offers inspiration for the archive that I treat in the chapters, but it also has implications for the critical practices appropriate to this archive, offering an opportunity to reconsider protocols of reading old and new. Mapping shifts in literary and cultural studies against this alternate periodization provides material for the crafting of a hybrid reading practice—a risk criticism— appropriate to our present age of environmental risk.

That new contexts might call for new practices is hardly news, but what the Clock might inspire us to ask is whether older practices might be retooled for new purposes. As a starting point for thinking about reading risk in the second nuclear age, I take science and technology studies (STS) scholar Bruno Latour's mobilization of the nuclear as metaphor, in this case of anachronism, in history fully as much as in critical practice: "After all," he argues, "masses of atomic missiles are transformed into a huge pile of junk once the question becomes how to defend against militants armed with box cutters or dirty bombs. Why would it not be the same with our critical arsenal, with the neutron bombs of deconstruction, with the missiles of discourse analysis?"[5] His point is precisely not to engage the nuclear, which is invoked only as the vehicle for the tenor of dangerous and destructive forms of critique, but his essay, which treats more directly the problem of climate change, perhaps inadvertently gestures toward a residual discourse that may in fact be amenable to his emergent practice, especially if we take his nuclear references as literally as his references to climate.

Apropos of the nuclear, Latour's are, of course, fighting words. In his metaphor, deconstruction *is* a neutron bomb, presumably leaving structures intact while taking the life out of them. What Latour does not engage directly, though, is that there was a more intimate relationship between the nuclear and those forms of critique for which he makes it stand. Those who inhabited the academy during the mid- to late 1980s might recall that deconstruction and discourse analysis were used, in fact, *against* such weaponry, in what was then (and sometimes still) called "nuclear criticism." Indeed, though Latour's targets in this essay are general and often moving—"we in the academy,"[6] "the social scientists,"[7] "the humanities"[8]—and he doesn't name many names, one cannot help but to read, behind references to deconstruction, discourse analysis, and even, at one point, "pharmakon,"[9] allusion to one of the foremost critics associated with bombs and deconstruction, Jacques Derrida. Derrida's seminal 1984 essay "No Apocalypse, Not

Now" set off a veritable chain reaction of poststructuralist accounts, with some critics suggesting that nuclear criticism might take its place among feminist criticism, Marxist criticism, and other established subfields of literary studies. Despite the persistence of the nuclear after the Cold War, however, the half-life of nuclear criticism seems to be of a shorter duration, with only a few of the most resilient critics persisting today.

In the contemporary era of environmental destruction, ecocriticism, the study of literature and the environment, might seem, quite rightly, to have taken nuclear criticism's place, and, given nuclear criticism's association with the Cold War, we may indeed wish to consider consigning elements of it to the dustbin of history. Though there were multiple nuclear criticisms, variously poststructuralist and ethico-political, all varieties were predicated on features of the atomic age that were fairly historically specific — the rhetoric of deterrence and the imagining of total thermonuclear war — both of which, in the age of dirty bombs and mininukes, might feel a bit anachronistic. When what the Clock measures is no longer only nuclear, but also chemical, biological, and atmospheric, the speeds are varied and the ends less sure. But, though some tenets of nuclear criticism are indeed likely "junk," I would like to experiment, following the imperatives of our age, with "reusing and recycling" aspects of nuclear criticism for a new era of risk that includes, but is not limited to, nuclear weapons and waste. Whatever its limitations, nuclear criticism had the merit of being a critical practice that grappled with the problem of the nuclear, that, in fact, suggested that literary studies, in particular, might be especially well suited to take such issues on, and in an era in which we are confronted with other large-scale risks of human origin, we might learn something from that earlier critical practice, even as we expose its historical blind spots, aporias, and constitutive omissions. My wager here is that bringing nuclear criticism and ecocriticism together under the rubric of something like a "risk criticism," a literary critical version of Ulrich Beck's risk society, might offer a way to theorize the megahazards of the present. And to do so in time — that is, in the risk temporalities of the second nuclear age.

NUCLEAR CRITICISM'S ENDS

As the Doomsday Clock suggests, time — and its end — had a key role in thinking the nuclear, in nuclear criticism fully as much as in nuclear popular culture. The discourses of deterrence, the notion of mutually assured destruction (or MAD), required the potential "midnight" of the Clock — that is, the possibility that there would be no future. This speculative orientation made the nuclear especially amenable to those critical practices with which Latour metaphorically associated it, deconstruction and discourse analysis, for it rendered the nuclear, as Derrida (in)famously put it, "fabulously textual," persistently present but only as so many models and imaginings. This total thermonuclear war is certainly the specter that haunts Derrida's "No Apocalypse, Not Now," in which the potential for a "remainderless cataclysm,"[10] "a total nuclear war, which, as a hypothesis, or, if you prefer, as a fantasy, or phantasm,"[11] provides the condition of possibility for nuclear criticism—and ultimately for the literature that such criticism might take as its object. Because the nuclear war to which deterrence narratives referred had not happened, except in text, it could have no real referent—only, as Derrida argued, a "signified referent."[12] Thus, the perpetual staging of that future event in the rhetoric of deterrence made nuclear war "fabulously textual"—though no less potentially hazardous as a consequence. Such textuality necessarily altered relations of expertise: because nuclear war has not occurred, no one is expert in it—all experts are working from speculative fictions (whether political or technoscientific)—and as readers of texts, literary and cultural critics are competent interpreters of the various representations of that fabulous event.

If this emphasis on the textuality of nuclear war seemed to authorize the expertise of the literary critic, Derrida went also a step further, suggesting that while the "remainderless cataclysm" could never be a real referent, it was also the ultimate referent, a referent conjured by the sign that marked the very limits of signification. Here, literature takes on a kind of analogical or homological relationship to the nuclear, for, if literature is defined, as Derrida suggests, as that which does not (as other discourses do) imply "reference to a real referent external to the archive itself," then this is something that it shares with the nuclear, which also "produc[es] and harbour[s] its own referent."[13] For Derrida, paradoxically, this "fabulous" referent is also "the only referent that is absolutely real" insofar as, if it were to come, it could not be recontained in the symbolic. Thus, while the "weaker" version of nuclear criticism applies the analytical tools of rhetorical analysis to the texts that figure the bomb, this "stronger" rationale makes nuclear war into a special instance of literature in general.[14]

With the nuclear end representing the possibility of a remainderless cataclysm, and literature representing that which can talk of nothing else, all literature becomes, in effect, nuclear literature, even when it does not thematize nuclear war and even when its publication precedes the nuclear age. Indeed, Derrida went so far as to tie "deconstruction" itself explicitly to the nuclear epoch. And other nuclear critics took up this association of textuality and the nuclear. Thus, Peter Schwenger, following on Derrida's observation that a nuclear war—with no one left to commemorate its purpose or memorialize its ideals—would be the first (and last) war in the name of the name alone, described the nuclear in terms of "an extreme example of the dominance of signifier over signified,"[15] his concern not with what literature might tell us about the nuclear but "what the nuclear referent could tell us about literature."[16] Similarly insistent on the symbolic power of the nuclear, William Chaloupka, alluding to the language of deterrence, asserted: "Never used but always effective, the power of the nuclearists could be seen as the greatest single accomplishment of the poststructuralist era."[17]

Nuclear criticism thus joined Cold War culture more generally in what Daniel Cordle has called a "state of suspense," predicated on an end that could have come at any time, and which, when it came, was to have been sudden, precipitous, and total. Nuclear criticism was therefore necessarily oriented toward the future, but in a way that also required imagining the future's nonexistence. The representation of time and the temporalities of representation are consequently central preoccupations of this work. As Kenneth Ruthven puts it, the instantaneousness of annihilation "destroys that slow-motion time-sense which our language mimes in the tense-system of its verbs, which separate out a past that was from a present that is and a future that will be."[18] This, according to Ruthven, is how one might account for Derrida's use of the future perfect in his "at the beginning there will have been speed"—"a nuclear beginning that will be simultaneously an end."[19] But this reading seems fairly imprecise, for the future perfect does not, in this case, accommodate the paradox of total thermonuclear war. Indeed, Richard Klein, commenting also on nuclear temporality, specifically rejects what he calls the "mimetic reassurance of a future anterior," in which "the future is envisaged as if it were the past": "Nuclear criticism denies itself that posthumous, apocalyptic perspective, with its pathos, its revelations, and its implicit reassurances."[20] If "there will have been," there must be a future time at which this will be true, which the total apocalypse-without-revelation of nuclear criticism disallowed. Klein indicated that what nuclear criticism might require by contrast is "a new, nonnarrative future tense," one that would avoid "the assumption that the future has a future,"[21] and he experimented with the paradoxes of the "Class A Blackout" and the "Prisoner's Dilemma"—both cases in which the future is predicated on a surprise that cannot be predicted—in order to grapple with this problem. In the case, then, of Derrida's "total war," Klein's prisoner's execution, or the *Bulletin*'s Doomsday Clock, the cataclysm is always to come.

In retrospect, however, that "fabulous" end seems not to have come.

The end of the Cold War and the dispersal of the referent-to-end-all-text called into question the utility of poststructuralist nuclear criticism. The focus on the textual qualities and future orientation of the bipolar nuclear conflict meant that nuclear critics to some extent colluded in the failure to recognize the multiplying effects of the nuclear on the ground. Nuclear critics tended to follow Derrida in saying that the bombs in Hiroshima and Nagasaki ended a conventional war rather than setting off a nuclear one, a distinction that safely kept the nuclear in the realm of fable.[22] Derrida clarified that no "non-localizable nuclear war" had occurred.[23] But "non-localizable nuclear war" might be a good way to characterize what did happen, as the United States, the Soviet Union, France, and others exploded those weapons throughout the Cold War period, and as the by-products of the nuclear weapons complex continue to plague us—potentially for millennia to come. As activists have long pointed out, the "fabulous" textuality that predominated in Cold War deterrence narratives always involved real explosions, nuclear tests that were to be read as signs pointing to that future annihilation. But as the real people, animals, and plants that were subjected to such tests knew, these weapons were no less real by having been treated as virtual. Of course, nuclear critics were not blind to the dangers of environmental peril, but the urgency of the fast apocalypse tended to eclipse that of the slow. As Schwenger put it:

> For most people the most disturbing fact about nuclear temporality is the instantaneousness of nuclear annihilation. If, as we are coming to understand, time is running out for the environment, time is at least still running. Nuclear disaster, on the other hand, is capable of occurring at any moment, in a moment, with no time even for an explanation of why there is no time.[24]

When the nuclear is only partially annihilating, however, the uniqueness of nuclear time—its instantaneousness, its surprise—diminishes, even as other risks multiply. Time is certainly still running, even as the disaster is also occurring at any (and every) moment.

TIME TO MOVE ON?

As the urgency of nuclear peril appeared to wane in the early 1990s, concern with environmental issues in literary studies grew, joining an environmental movement already very much in progress in the culture at large.

Though nature writing has, of course, been an important genre throughout American literary history, and attention to the place of nature in the cultural history of the United States characterized canonical works of criticism like Perry Miller's *Errand into the Wilderness* (1956) and *Nature's Nation* (1967) or Leo Marx's *Machine in the Garden* (1964), writing with an avowed interest in environmental politics tended to come from outside the discipline, as in exposés for a more general audience like Rachel Carson's *Silent Spring* (1962). By 1993, however, in an inversion of the order of priority outlined by Schwenger, Kenneth Ruthven was noting that environmental issues were eclipsing the nuclear, even among former "nuclear critics": "Our desire to forget about nuclearism is encouraged by the new environmentalists, who keep telling us that we have much more immediate things to worry about. Indeed, some of the latest doomsayers appear to have traded in their old CND [Campaign for Nuclear Disarmament] badges so as to begin campaigning on a green ticket."[25] Those nuclear critics whose interest had long been more conventionally ethical and activist (rather than more theoretical) in several cases did shift their attention to the environment, but often in the process left behind the nuclear.[26] In the fall 1991 issue of the newsletter for the International Society for the Study of Nuclear Texts and Contexts, for example, Daniel Zins opens his essay "Seventeen Minutes to Midnight" by recalling the response of a colleague to his workshop on "Environmental Security": "Daniel Zins—there's another one!" "What he meant," Zins explains, "was that here was yet another individual who, preoccupied with the problem of nuclear weapons during the 1980s, was now turning his attention to the possibility of *environmental* holocaust."[27] Indeed, by the next—and final—issue of the newsletter in the fall of 1992, the editor, Paul Brians, whose bibliography of nuclear texts provides an indispensable resource for the literature of the Cold War, was declaring "Farewell to the First Atomic Age": "The period originally called 'The Atomic Age' has passed: no more dreams of unlimited nuclear power, no more threat of nuclear ecocide. . . . It's time to move on."[28]

Naturally, such enthusiastic postmortems were rather premature. What was then anachronistic was not the nuclear per se but the end-times with which it had been associated. With the dispersal of nuclear risk, the chance of *total* thermonuclear war—the Doomsday Clock's midnight—diminished, and this was, as it turned out, fairly fatal to the nuclear criticism imagined by critics like Derrida. Meanwhile, ecocriticism emerged as an upstart counter to poststructuralism—and to Theory more generally. Emphasizing the persistent "real" referent of present and continuing environmental damage, early ecocritics, practitioners of what Lawrence Buell has called the "first wave," "looked to the movement chiefly as a way of 'rescuing' literature from the distantiations of reader from text and text from world"[29]—hardly the discourse most amenable to those whose concern had been with the "fabulously textual." If what interested poststructuralists about the nuclear was the impossibility of representation, ecocritics by contrast seemed to be returning to "what superficially seems an old-fashioned propensity for 'realistic' modes of representation."[30] In this way, first-wave ecocritics had some affinities with those nuclear critics who were driven more by an ethico-political commitment to disarmament than by an interest in the philosophical issues raised by total annihilation; these more traditionally humanist scholars shared what Daniel Cordle describes as a "very old assumption in English studies: books are, in a rather nebulous sense, good for you."[31] One reads speculative fiction of nuclear holocaust in order to learn how to avoid it; one reads nature writing in order to learn how to be more ecological.[32]

Ecocriticism's privileging of the real was an important corrective to what did at times get a bit "loony"—to borrow Christopher Norris's term—in poststructuralist-inspired work. Responding to those critics who took Derrida's "there is no outside-the-text" too literally, Norris asserts that "it is time to enounce a few simple truths that literary theorists seem bent upon forgetting," including that "textuality doesn't go 'all the way down.'"[33] Or, as Kate Soper noted in another context, "It is not language that has a hole in its ozone layer."[34] Of course, as subsequent "second wave" ecocritics (Soper among them) have pointed out, it is just as clear that how we know there is an ozone hole and what we think we ought to do about it are products of mediation and discursive production. As a consequence, ecocritical approaches that highlight "hybridity" or perform discourse analysis or emphasize the importance of text have increasingly supplanted what might initially have been an overinvestment in access to the real. But even as some ecocritics have returned to poststructuralism to mine it for latent ecologies, few have excavated the environmental potential in poststructuralism's nuclear turn.[35]

If we accept the *Bulletin of the Atomic Scientists*'s contention that what we currently inhabit is something like a second nuclear age, we might require a revamped nuclear criticism to account for it, and Daniel Cordle has made a convincing call for a nuclear criticism today that might usefully historicize the exceptionalist rhetoric associated with the War on Terror.[36] Indeed, we might take Paul Brians's spatial metaphor literally: In the "second nuclear age" it is indeed "time to move on," not to other issues but to other locales—India, Pakistan, Iran, North Korea—and even to other, related, nuclear concerns—nuclear power and waste, uranium mining, depleted uranium weaponry. As Roger Luckhurst noted, by the time of the publication of his own Derrida-inspired essay in *Diacritics* in 1993, nuclear criticism had become, not *anachronistic*, but *anachoristic*, confronting "misplacements, 'unreadable' geo-graphical loci" produced "on the terrain of dissolved Cold War certainties."[37] And, as though in response to nuclear critic Ken Ruthven's admonition that "if it is to engage with the late twentieth century . . . , the new ecocriticism can hardly avoid being contaminated by the concerns of nuclear criticism,"[38] ecocritics influenced by the environmental justice movement have continued to address ongoing issues of nuclear politics, in the Pacific and elsewhere.[39] But as productively anachronistic as calling such practices "nuclear criticism" might be, I would submit that the literary criticism of the second nuclear age ought not to make the mistake of the first in fetishizing the nuclear over other megarisks. Rather, I would suggest that there are, within ecocriticism, nuclear criticism, and sociological theories of risk, elements of an emergent risk criticism, an umbrella practice that might accommodate both the "fabulously textual" and the absolutely material qualities of contemporary risk.[40]

RISK CRITICISM

If what transformed questions of expertise and competence in the "first nuclear age" was, as Norris shorthands it, that "self-appointed 'experts' are so manifestly out of their depth—confronting such a range of intractable problems, aporias, or wholly unforeseeable turns of event,"[41] this is surely doubly true in the "second nuclear age," when climate change and biotechnology have been added to the catastrophic mix. In this context, perhaps no contemporary theorist has dwelt so singularly on questions of hazard, expertise, and representation as Ulrich Beck, who has, from his inaugural study, *Risk Society: Toward a New Modernity* (1986), to his more recent iteration, *World at Risk* (2007), persistently queried the conditions of what he calls "world risk society," a period of "reflexive modernity" in which the unanticipated (and unanticipatable) side effects of technological innovations come back to haunt us. To use Beck as a guide is, in effect, already to take a stance, for Beck's "risk society" is, like Derrida's nuclear criticism, a critique rather than an endorsement of the discourses of risk management intrinsic to both. In this way, risk is a misnomer, for many of the specific cases that Beck cites are examples of the failures of risk prognostication, those hazards, or even full-blown ecocatastrophes, of the twentieth and twenty-first centuries, from Chernobyl to Fukushima. Risk is nonetheless a useful watchword—for Beck's project and for mine—not only because it references the dominant discourse of "management," but also because it highlights precisely what cannot be managed, the incalculable effects of risky technologies.

As a sociologist, Beck is clearly not himself a risk assessor or manager, but his approach is also importantly different from the seemingly more compatible approaches to the cultural perception of risk offered by scholars of "risk communication"—those who focus on the ways in which risk is communicated to a lay public—whose pretense to neutrality can translate as a sympathy for the expert or risk manager's predicament. The focus in such work on the "stigmatizing" of risk, for example, which William Leiss and Douglas Powell associate with dioxin, and Jeanne X. Kasperson and coauthors extend to the nuclear industry, Alar in apples, Tylenol contamination, and the oil industry, may give the impression that such hazards are simply misunderstood by an uninformed lay public, an impression heightened when Kasperson and coauthors liken such stigmatizing to the *A* in Hawthorne's *The Scarlet Letter* or the yellow star in Nazi Germany.[42] Such would seem to confirm that there is indeed an "in-built anti-environmentalist bias" in the discourse.[43] As Ursula Heise notes, however, this version of the "social amplification of risk" has been joined by a discourse more critical of risk "experts," as in Charles Perrow's work on the "normal accidents" intrinsic to high-risk technologies. The differences between these approaches to risk might be represented in Perrow's and Paul Slovic's disparate responses to the Fukushima accident, both published in the *Bulletin of the Atomic Scientists*. Slovic focuses on what he calls a "perception gap" on the part of a fearful public, a gap remedied by informing laypeople of the other myriad sources of radiation exposure, from plane flights to radon. The cautionary tale of Fukushima becomes a story of preemptive risk communication: "Messages should be created and tested before the next emergency," he warns, for "if they are not, the next disaster response will, in hindsight, cast a harsh light on officials who failed to prepare for the known communication challenges."[44] For Perrow, focusing on the technology itself rather than on the communication of its risk, however, the lesson is rather that "some complex systems with catastrophic potential are just too dangerous to exist, not because we do not want to make them safe, but because as so much experience has shown, we simply cannot."[45] Who or what is at risk is thus quite different here—for Slovic, it is the officials at risk of looking underprepared; for Perrow, it is the rest of us.

In the face of the very real hazards of such technologies, and the very real problems with risk management discourse, one might wish to use terms like "hazard" or "danger" in place of "risk," but to make such substitutions is not only to avoid confrontation with the dominant discourse of risk management but potentially also to elide the real rhetorical and political difficulties with navigating true uncertainty, when we suspect but don't know, or don't know in a way that is legible to those in charge. Emphasizing the speculative qualities of risk, Beck offers what risk critics might interpret as a kind of vocation for literary studies: Risk is virtual, and, "without techniques of visualization, without symbolic forms, without mass media, etc., risks are nothing at all."[46] If theorists of the "social amplification of risk" would focus on the challenges of defusing lay misconceptions, in the process implying certainty about the relatively benign nature of the technology in question, risk critics like Beck, less sanguine about expert assurances, take a more critical approach to discourses of risk, taking as seriously the staging of risk by laypeople and activists as that by technicians, actuaries, and risk managers.

Risk society is global and, like the nuclear age, oriented in part toward the future. As in the *Bulletin's* "second nuclear age," Beck's work places the nuclear among other "megarisks" of the late twentieth—and now early twenty-first—century, though he sets the origins of this "age" somewhat earlier, somewhere in the 1960s or 1970s. For Beck, genetic engineering, ozone depletion and global warming, chemical and toxicological contaminations, *and* nuclear threats all mark a shift from what he calls the "first modernity" of industrial society to the "second modernity" of global risk. And though he does acknowledge the unevenness of global risk distribution—with some benefiting from others' losses; some able to shield themselves from hazard, environmental or otherwise—he perhaps optimistically anticipates that risk society will also be characterized by a kind of cosmopolitan spirit of shared hazard, as a common risk makes for a common bond. Such cosmopolitanism requires, though, representation through which global citizens of world risk society might come to experience their shared condition of "being-at-risk." Like nuclear criticism, risk criticism confronts a willingness to wager the future in a high-stakes game, whether of nuclear deterrence or technological fix. But if the signature fact of nuclear criticism's "remainderless cataclysm" is its remainderlessness—with no one to mourn, to recontain the event in the symbolic—in the case of risk, Beck suggests that "we are . . . experiencing . . . the fact that *people have to keep on living afterward.*"[47] Survival necessarily challenges the temporality of threat—and indeed the very nature of the catastrophic event.

In the case of contaminations by radiation or chemicals, of rising sea levels or potential allergens in genetically modified foods, we are in a situation that seems to imply the inverse of Derrida's argument regarding the "competence" of the humanities in general and literary studies in particular. In the case of the "fabulously textual" war, there was no "real referent" about which any of the so-called experts might claim to have expertise. In the case of risk, by contrast, one must posit the existence of a "real referent," and access to the means of its representation appears indeed to be restricted to technoscientific experts and government regulators. Here, the question shifts from a contemplation of one's future nonexistence to a contemplation of the nature of one's continued existence, and, as Peter van Wyck has pointed out, "Am I already a casualty? is exactly the question provoked."[48] Describing the particularly elusive case of radiation, van Wyck continues:

> It is as though our senses, our very perception, had been expropriated, rendered useless and vestigial in the face of threats that cannot be seen, heard, smelled, tasted, or touched. The appeal to the eyewitness (even one's own eyes) comes to have little value here. There is nothing there, nothing to be seen, leaving us dependent on others (often the same others, that is, the same institutions that produced the threats) to determine the appropriate means (instrumentation) with which to represent it safely back to us and for us.[49]

Environmental risk would thus seem to place laypeople, including humanists and literary critics, in a balefully weakened position, waiting for the experts to translate the real that may already be causing harm.

Such, at least, is the official line of what, for Beck, is "first modernity," a world in which experts approach a calculable future. As Beck points out, however, risk assessments are often wrong—and catastrophically so. From CFCs and the ozone hole to asbestos and cancer, from thalidomide to BPA, "New knowledge can transform normality into threat overnight." In the face of shifting expert opinions, recalls, public apologies, "The progress of science refutes its original security assurances," "sow[ing] the seeds of doubt concerning its declarations about risk."[50] Scientific risk assessment would attempt to construct a narrative of present cause and future effect, but, as Beck notes, the limits of such predictions are likely clear to anyone who follows the news on a regular basis, as "what was judged 'safe' to swallow today, may be a 'cancer risk' in two years' time"[51]—or, rather, in what may well be the key tense of risk society, *will have been* a cancer risk, in a way impossible to predict in advance. Here, Klein's "reassurance of the future anterior" seems less than reassuring. More "knowledge" produces less certainty, and risk calculation produces incalculable risk.

As nuclear criticism was an analysis of the illogic and limitations of what used to be called "nukespeak," so Beck's risk analysis is a response to risk discourse. As van Wyck notes, risk "is but a neologism of the insurance industry," and, in risk society, the limits of "thinkability" are replaced by the limits of insurability.[52] These latter limits, posed by risky technologies old and new, may shed new light on another essay that appeared alongside Derrida's in the special issue of *Diacritics*. In an analysis of the nuclear sublime, Frances Ferguson opens with a quote from a letter she received from the State Farm Insurance Company: "Under no circumstances does your policy provide coverage for loss involving nuclear accident."[53] For Ferguson, this limit is one of imagination—the failure to think the unthinkable, as nuclear war was often called—but nuclear accident is not necessarily the nuclear holocaust with which most nuclear critics equated it. The State Farm notice, unlike Derrida's cataclysm or Klein's nonnarrative future tense, does presume a future, a moment after which the nuclear event will have happened—and will have been an event; it just presumes a future for which the insurance company would not like to be liable. This is the point at which risk-benefit analysis breaks down; at which the "maximum credible accident" produces losses that can be imagined, just not compensated.

For Beck, counterintuitively, risk is also in effect "fabulously textual" insofar as it is not visible until it is represented—or, as Beck prefers, "staged." Unlike nuclear war, however, that transition from signified to real referent is ongoing. As Beck puts it, risk is always in the process of "becoming real."[54] A paradigmatic example of this process comes in Peter van Wyck's book *Signs of Danger*, in which he describes the case of a Soviet film documenting the aftermath of the Chernobyl accident. *Chernobyl: Chronicle of Difficult Weeks* is, van Wyck notes, "a clumsy piece of backslapping propaganda showing how well the Soviet scientific, technical, military and party authorities came together in the face of great adversity to overcome the severity of the accident. . . . But what we see on the surface of the film itself," he continues, "are millions of tiny pops and scratches."[55] The filmmakers initially presumed that they had simply inadvertently used defective film stock, but, van Wyck explains, what they had in fact represented was "a record of the impacts of decay particles as they passed through the body of the camera . . . a very striking pointillism of the real—discovered only after the fact, only retroactively."[56] The "signs of danger" here do not signify until they are read, and this reading in turn makes legible the limits of the "expert" narrative. The fabulous textuality and the absolute materiality of risk are represented quite graphically in the medium that now must become also the message.

Thus, as in "No Apocalypse, Not Now," literary and cultural critics are not just competent to take on the technoscientific and political issues posed by risk but are in some sense particularly qualified, as readers of the symbolic forms by which risk becomes real. And, inspired by Beck's work, several literary critics have taken up the task of risk discourse analysis. Referring to what he calls "toxic discourse," Lawrence Buell reads across fictional and nonfictional representations of toxicity to excavate its discursive roots in the gothic and the pastoral.[57] And, Frederick Buell reads a number of literary texts in the final chapter of his analysis of dwelling in apocalypse, *From Apocalypse to Way of Life* (2004). Extending this work, Ursula Heise includes a wide-ranging discussion of risk in her book *Sense of Place and Sense of Planet* (2008), in which she turns to contemporary novels that thematize risk in order to explore the effect these themes might have on narrative. These approaches shed new light on such much-read texts as Don DeLillo's novel *White Noise* (1985), which emerges less as an example of postmodern satire than as the paradigm of risk realism, as the "hyperboles and simulations that have typically been read as examples of postmodern inauthenticity become . . . manifestations of daily encounters with risks whose reality, scope, and consequences cannot be assessed with certainty."[58]

As the work of these critics suggests, risk society also may offer us a somewhat different model for "literature" than the one envisioned in the first atomic age. For Derrida, literature shared its "fabulous textuality" with nuclear war—both produced and contained their referents, and in this way remained sealed off from the real, even as they might have produced material effects. In retrospect, though, the idea that the nuclear "produces and harbours" its own referent looks a bit like its own sort of containment strategy, a way for the nuclear critic to remain suspended in textuality at the expense of materiality. In an age of leaks and spills, such quarantining of literature may no longer be tenable either. Indeed, if we needed an iconic example of the ways in which risk is perpetually moving from virtual to real, we need look no further than *White Noise*, which seemed to anticipate the Bhopal explosion in India, or Beck's *Risk Society*, the publication of which coincided with the accident at Chernobyl. For risk narratives, anticipation may already be description.

TURN BACK THE CLOCK?

Prognostication is a tricky practice, and we might be tempted to leave it to the experts like those at the *Bulletin of the Atomic Scientists* who at least are motivated not by limiting liability but by a desire to assess material threats. By their metric, that icon of risk communication, the Doomsday Clock, the second nuclear age is a risky place. In the years since 2007, the *Bulletin* has moved the Clock several times. In 2010, with much fanfare, including a live webcast of the event at TurnBackTheClock.org, the board moved the Doomsday Clock a cautiously optimistic one minute further from midnight (from five to six). Clearly eager both to acknowledge the significance of the early days of the Obama administration and to raise awareness of and action on world risk in a moment in which crises in the economy seem to be eclipsing crises of ecology, the editors explained: "By shifting the hand back from midnight by only one additional minute, we emphasize how much needs to be accomplished, while at the same time recognizing the signs of [international] collaboration."[59] Thus, acknowledgments of dire risk ("our habitat could be disrupted beyond recognition" by climate change; nuclear power generation could lead to nuclear weapons development; advances in the life sciences pose risks, "either inadvertent or intentional") are tempered by hope (the—very—modest accomplishments of Copenhagen; the promises of alternative energy). But a mere two years later, with slow progress on climate change and proliferation alike, the Clock was set back to five minutes to midnight, and most recently, in early 2015, the Clock was moved two minutes further still, such that it stands, at this writing, at a mere three minutes to midnight, the same setting it had in 1984 and 1949, and only one minute away from its closest setting to date. The editors concluded these announcements with the ominous reminder that has long accompanied the clock's readjustments: "The clock is ticking."

As a risk critic, an expert, not in nuclear physics, toxicology, or climatology, but in literary and cultural analysis, my strategy here has also been (in the hopeful spirit of the *Bulletin*'s activist admonition) to "turn back the clock"—in this case, to measure the distance from Cold War nuclear criticism to post–Cold War risk, and to ask whether elements of that earlier practice of the first nuclear age might usefully be retooled for the second. But I would not be doing my job if I did not also turn to the Clock itself as an object of analysis. Here, the *Bulletin*'s Clock provides a useful corrective to any narrative of progress that might mislead us into complacency regarding the nuclear, and with the other catastrophic dangers pushing the minute hand toward midnight, the call to "turn back the clock" seems pressingly urgent. But broadening the source of risk beyond the nuclear raises questions about how far back the Clock need go. The second nuclear age usefully turns our attention to the other megarisks of the contemporary moment, but the origins of *those* crises clearly predate 2007. If we are to take seriously the conjunction of concerns organized under the category of environmental hazard, then surely at minimum Beck's risk society, with its origins in the 1960s, or even the first nuclear age in 1945, would be better frames of historical reference. Indeed, in a context in which some historians, like Dipesh Chakrabarty, have begun to identify the present as part of the Anthopocene—the era in which "humans act as a main determinate in the environment of the planet" (roughly 1750 to the present, according to his sources)[60]—the *Bulletin*'s recognition of other environmental threats seems decidedly belated.

And if the Clock itself needs changing—its face measuring different scales of the temporality (and, indeed, spatiality) of risk—its ends too must change. The iconic nuclear midnight signifies differently when other threats are included, for, referring both to the potential for sudden and total annihilation—that fear that so dominated the Cold War—and to what Andrew McMurry calls the "slow apocalypse," midnight begins to look as though it may have been the wrong metaphor all along, for "by holding out for that noisy demise, we can pretend we haven't been expiring by inches for decades."[61] "Experience of this sort," Frederick Buell explains, "does not involve a terrible and conclusive moment ahead when people breach nature's limits and disrupt nature's fundamental equilibrium; it means that one has already entered (or perhaps is already well into) a time when limits have been breached and the risks from disequilibrium are rising."[62] In this sense, what the Clock must measure is indeed, as Derrida suggested of the world post-9/11, "worse than the Cold War," a threat that "represents the residual consequence of *both* the Cold War *and* the passage beyond the Cold War."[63] The Clock might better be conceived, then, not only as a countdown to some potential future annihilation, but also as what Peter van Wyck calls a "metronome of threat,"[64] syncopating a present that contains multiple catastrophes, historical and to come, simultaneously. In the second nuclear age, we seem to find ourselves inhabiting rather than anticipating the end, as though, as one of the characters in Lydia Millet's nuclear novel, *Oh Pure and Radiant Heart* (2003), puts it, "The end has already come and gone. And here we are."[65]

When Derrida said, "no apocalypse, not now," he meant no revelation, even in the case of what might colloquially be considered apocalyptic, a total thermonuclear war. The nuclear epoch was thus, he argued, also an épochè, "suspending judgment before the absolute decision."[66] In the present age of risk, we are clearly no closer to "absolute knowledge," but though we arguably persist in a "state of suspense," we do not await the same end. Turning back the Doomsday Clock has always entailed turning it forward, imagining the future so as to avert catastrophe. In a time in which more knowledge seems to create less certainty, as the "becoming-real" of risk offers a kind of ongoing revelation without foreseeable end, we need to be as aware of the past and present as we are oriented toward what might come. But, still, the precautionary principle implied in the future anterior remains indispensable. In a dialogue with Giovanna Borradori following 9/11, even while projecting once again "a future so radically to come that it resists even the grammar of the future anterior," Derrida himself produced just such a speculative fiction: "One day it might be said: 'September 11'—those were the ('good') old days of the last war. Things were still of the order of the gigantic: visible and enormous! What size, what height! There has been worse since." And, extrapolating one example of the "worse," Derrida imagined, "nanotechnologies of all sorts [that] are so much more powerful and invisible, uncontrollable, capable of creeping in everywhere."[67]

As the apocalypse "creeps in," the problem is less unthinkable nonexistence, the absolute end in a remainderless cataclysm, than survival in a world that Derrida described as "autoimmune," turned inside out, as the "environment" becomes what Bruno Latour would call a hybrid, anthropogenic and ecological, discursive and real, that we can neither predict or control. It is telling, in this context, to compare the section headings in Derrida's Cold War–era "No Apocalypse, Not Now" and his post-9/11 "Autoimmunity: Real and Symbolic Suicides": in the former, he describes his thoughts as "tiny inoffensive missiles," a rhetoric of sending and receiving with its inevitable failures and misfirings; in the latter, these sendings become internal processes of "reflex and reflection." As in Beck's "reflexive modernity," this is not a moment that is newly self-conscious or thoughtful, though that could be a consequence and is doubtless what a risk critic might hope to achieve, but a society confronted by itself, raising the question (and the stakes) of what sort of critical stance might be appropriate for the shifting ground of risks in the process of becoming-real.

PRECAUTIONARY READING, OR "MAKING UNCERTAINTY A PRIMARY ISSUE"

Risk criticism in the second nuclear age negotiates an uncertain terrain between the speculative and the real, the risk, the hazard, and the catastrophe. Indeed, the real was very much the concern for Bruno Latour when he worried about the dangers of wielding an outdated arsenal of critical practices, the "neutron bombs of deconstruction," the "missiles of discourse analysis," for part of what inspired his apprehension was an editorial titled "Environmental Word Games," which he read in the *New York Times* in 2003. This artifact of the bad old days of the George W. Bush administration makes the by now hardly surprising assertion that, rather than producing a more environmentally friendly policy, that administration worked to produce a friendlier *sounding* policy, to "dress it up with warm and fuzzy words," even as it injected doubt in the scientific veracity of environmental claims—an explicit and targeted effort, as Republican strategist Frank Luntz then explained, to "continue to make the *lack of scientific certainty* a primary issue."[68] This strategy plays on the cautiousness of science, on the precariousness of modeling complex systems, on the diversity of interpretation, and it has been thoroughly exploited by dissenters to the "theory" of climate change. But this strategy is worrying for Latour because he finds it so uncannily familiar. Though his own motivations were opposite those of Luntz, Latour himself has "spent some time in the past trying to show 'the lack of scientific certainty' inherent in the construction of facts," even if his intent was to *"emancipate* the public" rather than to obscure a clear and present danger.[69]

Faced with the need to grapple with the reality of global warming, critics must delineate science fiction from science fact. The very habitability of the planet seems to require it. And those older tools, "the neutron bombs of deconstruction" and "the missiles of discourse analysis," tools that are often used indiscriminately on a variety of literary and nonliterary texts, tools that excavate and interrogate truth-claims, often for the purpose of "mak[ing] the lack of . . . certainty a primary issue," may not suffice. Given these conditions, for Latour, "The danger would no longer be coming from an excessive confidence in ideological arguments posturing as matters of fact—as we have learned to combat so efficiently in the past—but from an excessive *distrust* of good matters of fact disguised as bad ideological biases!"[70] But neither is Latour willing simply to accept the firm certainty of scientific "fact," unmoored from consideration of historical context and scientific practice. We need not simply accept the "truth" handed down from the technoscientific experts, especially since, as Beck argues, "The 'truth' changes."[71] Thus, reprising an argument he has made repeatedly (at least since *We Have Never Been Modern* [1991]), Latour asserts that both cultural constructionist and realist models of critique are emblems of a "debunking" critical stance that has, he argues, "run out of steam." For Latour, the objects of investigation were always constructed by discourse, produced by material practices, and themselves animate agents, transforming the world in turn. Supplementing Derrida, we might argue that global warming is at once "fabulously textual" and absolutely material, a product of expert assessment, media presentation, political accord, and public reception, as much as it is an interaction of CO_2 and methane gases in the atmosphere. Climate is not, then, a matter of fact, in the positivist sense of that phrase, but, borrowing from Latour, a "matter of concern," a gathering of diverse actants that together coproduce its reality.

But as sure as we may be about the reality of climate change (or radiation danger or the effects of toxic chemicals), risk means also taking seriously "lack of certainty." Uncertainty *is*, in risk society, a "primary issue," bringing together the unlikeliest of bedfellows. Surely on no other topic would such critical luminaries as Ulrich Beck and Slavoj Žižek both cite former U.S. secretary of defense Donald Rumsfeld, whose comments on uncertainty during the buildup to the Iraq war have become iconic. "As we know," Rumsfeld famously said, "there are known knowns; there are things we know we know. We also know there are known unknowns; that is to say we know there are some things we do not know. But there are also unknown unknowns—the ones we don't know we don't know."[72] In context, of course, Rumsfeld was making an argument for preemption, for proceeding in the face of uncertainty, a move quite contrary to that other Republican strategist Luntz, who seems to imply that uncertainty underwrites inaction. Yet another strategy gaining momentum today is the highly ambivalent concept of "resilience," a term that appeals both to beneficiaries of what Naomi Klein called "disaster capitalism" and potentially also to ecologists and activists who might wish to envision a world of salutary adaptations. But there is another option here—precaution, a principle that undergirds much environmental thought. Uncertainty is central, for risk critics fully as much as for Republican strategists; the question is what to do in the face of it.

Precaution has emerged as a response to risk, offering an alternative to science-based risk assessment that, while also focused on the future, emphasizes its uncertainness, unpredictability, and incalculability, and a precautionary reading practice might offer a mode for risk critics, interested both in tracking the ethico-political consequences of risk and potentially in forestalling ecocatastrophe. Precautionary reading might operate as an extension of Beck's risk isomorphism, the suggestion that megarisks share a similar shape that renders them recognizable as hallmarks of a risk society. Commenting, for example, on the use of the precautionary principle as a response to genetically modified foods, Kerry Whiteside acknowledges that "there has been no biotech Chernobyl."[73] But his implication is that there could be, that despite the clear differences that separate nuclear and biotech, these issues are structurally similar, and thus precaution ought to be mobilized as a kind of speculative analogizing, and one that carries with it an ethics of accountability and responsibility. As François Ewald notes, "Under the old approach to responsibility uncertainty of knowledge was innocence"; thus, lack of scientific certainty of harm meant lack of responsibility when the harm came to pass. In the precautionary regime, responsibility shifts and one is judged, Ewald says, "not only by what one should know but also by what one should have or might have suspected."[74] We cannot know the future consequences of our actions, but we will nonetheless have been responsible for them.

Precaution is in part a reading practice and in part a speculative fiction, oriented backward to past catastrophes and forward to potential ones. And, though the precautionary principle is sometimes indicted for its conservatism, its preference for the more comfortable continuance of business as usual, this is only one potential consequence of precautionary thinking. Contrary to those who would argue that precaution is only and always a brake on innovation, Latour clarifies: it "does not simply mean that we stop taking action until we are certain about the innocuousness of a good, for that would once again return us to the ideal of mastery and knowledge by demanding certain knowledge about an innovation which, by definition, and like any technology, forever escapes mastery";[75] rather, the principle of precaution is, as Kerry Whiteside usefully puts it, "uncertainty made conscious," a critical task surely as appropriate to literary studies as it is to science, governance, and industry.[76] In this context, literary and cultural critics might take inspiration from Kerry Whiteside's description of the obligations posed in the principle of precaution:

> Precaution obliges us to examine and discuss the innumerable linkages through which we come to know nature. It diversifies the contacts at the interface of nature and humanity. It problematizes and opens to discussion the values that are implicit in the scientific framing of environmental issues.[77]

Discussing linkages, diversifying contacts, problematizing values—these are the practices in which we might claim some expertise. And if precaution acknowledges that we cannot control the future, this is a lesson taught as well in the *Bulletin*'s Clock, hardly the icon of technophobic reactionaries.

The strategy of the Clock is the strategy of the future anterior—we have never been at midnight, so we cannot in fact know our proximity to the end. We will have been at three minutes to midnight; whether we are there now is an act of speculative imagining, whatever sort of scientific calculations might be involved. Such doomsday apocalypticism has come in for some critique, for, numbed by proliferating environmental crises, we carry on with business as usual. But the techno-optimism of the open future, in which presently developing calamities will be remedied by the innovative pluck of our technosavvy grandchildren, has something of the "old woman who swallowed a fly" (and then a spider and a rat and a cat . . .) about it. After all, our grandchildren might remind us of the refrain of that ditty: "Perhaps she'll die."[78] Writing of the use of the future anterior in Baidou, Žižek argues that the "circular strategy of the *futur antérieur* is also the only truly effective one in the face of a calamity (say, of an ecological disaster): instead of saying 'the future is still open, we still have the time to act and prevent the worst,' one should accept the catastrophe as inevitable, and then act to retroactively undo what is already 'written in the stars' as our destiny."[79] The future anterior, as the Doomsday Clock reminds us, is not really about the future; it is about the present as someone's past. The threats of the present are in some ways analogous to those of the first nuclear age, for they too are unprecedented; expert knowledge is often speculative fiction attempting to resolve the present uncertainty, which remains a primary issue, even as that does not absolve us of responsibility. The nonlocalizable nuclear war of the Cold War period was a kind of side effect of the fabulous war to come; the political imperative of the second nuclear age is to recognize side effects as effects, collateral damage as violence, whether that comes in the bombing of civilians, in the precipitous meltdown of a reactor, or in what Rob Nixon so eloquently calls the "slow violence" of environmental injustice.[80]

It is in this spirit that *Risk Criticism* aims to codify and extend work in literary and cultural studies that grapples with contemporary risk, in the process inevitably grappling with the problems of containment and the limits of expertise. Though the primary terrain of the book is North American, it is in the nature of risk to resist any such geographical boundaries. At the same time, while the globalized nature of risk can be an opportunity for global solidarity—an internationalism that characterized the early atomic scientists' warning that the bomb meant "one world or none"—I am mindful of the fact that risks are always represented and experienced unevenly. The potential archive of risk is, clearly, capacious; my intent here is to be more explorative than exhaustive. Interested both in tracking and intervening in the ways in which risk is represented among laypeople experiencing the human condition of "being-at-risk," I focus primarily on literary texts—fiction, poetry, drama—but I also bring a number of nonliterary texts to bear—from visual art (photography, painting) to popular science and documentary films—and I draw on the expertise of scholarly work from across the environmental humanities. My objects of inquiry are at once fabulously textual and absolutely material, but I am first and foremost a literary critic, trained in parsing, not contaminants, but texts. Representing global risk inevitably involves what Hayden White once called "tropological wagers," those linguistic strategies that underlie the writing of history, which, borrowing from Kenneth Burke's notion of "master tropes," White identifies as metaphor, metonymy, synecdoche, and irony. For White, historiography might identify which of those tropes animates any particular work of history. Risk discourse, too, is characterized by wagers; the practice of risk criticism is to excavate those tropes, to explore their implications and potential side effects.

Risk Criticism takes up some of the tropological, theoretical, and material concerns central to the synecdoche that is the second nuclear age. Chapter 1 addresses more directly the question of the archive risk criticism might take as its object. Building on questions of temporality, I take up Beck's suggestion that the dominant trope of risk society is irony, as modernity in general and the sciences in particular become "reflexive," confronted with catastrophes of their own making. As a case study for considering questions of expertise and responsibility in the second nuclear age, I turn to the specter of the atomic scientist, a figure that, as Foucault argued with reference to Oppenheimer, seemed to transcend the particularity of disciplinary training and scholarly divisions of labor. Here, I read portions of the rich archive that the atomic scientists themselves produced, including selections from the anthology *One World or None* (1946), as well as works by J. Robert Oppenheimer and Leo Szilard. Both scientists wrote essays on science, politics, and ethics for a general audience, and Szilard also published a book of science fiction, *Voice of the Dolphins* (1961), on the subject. In the spirit of reflexivity, I read this work of the early atomic age alongside literary texts that highlight what Beck calls the "involuntary satire" of global risk: Kurt Vonnegut's novel *Cat's Cradle* (1962) and Toni Cade Bambara's *The Salt Eaters* (1980), both of which provide very different models of the scientist as public intellectual, and Lydia Millet's more recent novel *Oh Pure and Radiant Heart* (2005), in which Oppenheimer, Szilard, and Enrico Fermi are magically transported from 1945 to 2003, where they attempt both to come to terms with the world their "gadget" has wrought and to remount—first time tragedy, second time farce—an ultimately unsuccessful "global" resistance movement.

Chapter 2 addresses the consumer and agricultural products that chemical companies offered following the war (often the same companies that were deeply invested in the technology and production of the atomic bomb). Focusing particularly on the Union Carbide accident in Bhopal, India, I frame this chapter with a rhetorical move common to thinking "global" risk—a kind of metonymy mobilized as metaphor—evinced in the title of a short essay published following the accident: "We all live in Bhopal." Pointing to the ambivalence of this strategy, I turn to Don DeLillo's *White Noise* (1985), a text that, in the budding discourse on risk in literary studies, has become a kind of touchstone. Common in these readings of the novel is what I call the "Bhopal gesture," a tendency to reference the accident as evidence of DeLillo's prescience in writing of an "airborne toxic event." Useful as this comparison can be, it also can have the effect of effacing the particularities of the U.S. context, in which farmworker movements were quite vocal in trying to bridge the conceptual and geographical distances between producer and consumer when it came to pesticide risks— the glimmering awareness of which I read in the grocery story sections of *White Noise*. Arguing that the "reduction" of metonymy might be replaced by a principle of addition, I conclude the chapter by pursuing the Bhopal comparison more directly by putting DeLillo's novel in conversation with Indra Sinha's *Animal's People* (2008), a novel that highlights the particularities of "living in Bhopal," even as it calls for a cosmopolitan coalitional response to this and other environmental catastrophes.

Moving from known (if contested) risks into the realms of less well established potential risks, chapter 3 treats controversies surrounding genetic modification, the technology that is often touted as "solving" a range of problems, including the use of toxic agricultural chemicals. Arguing that the predominant emotion accompanying new technologies like GMOs is less "fear" (which Beck highlights) than "discomfort," I turn to the uncertainties that generate these emotions: Are genetically modified foods "Frankenfoods"? Are they benign or malign? Will they transform all life on the planet? To what extent are they, as the regulatory discourse of "substantial equivalence" would have it, analogous to conventional foods? Though there have been speculative and science fictional attempts to answer these questions, I focus here instead on the uncertain present, turning to a realist attempt to grapple with the "becoming-real" of GM risk: Ruth Ozeki's novel *All Over Creation* (2003). Ozeki's novel targets the company that has come to stand, metonymically, for biotech, Monsanto, and represents nonexpert attempts to unveil the dangers hidden in these products, opening up the potential for nonequivalence in the regulatory analogizing. The novel closes with a gesture toward the global reach of Monsanto, both in Ozeki's imagined corporation's new marketing plan and in the Seattle World Trade Organization protest in 1999. Carrying this narrative forward, I put *All Over Creation* in conversation with a public debate, staged by Google's philanthropic arm, Google.org, between writer/activist/academic Michael Pollan and Monsanto's CEO Hugh Grant, in which the analogue for life becomes data—both to be engineered and to be stored (or backed up) in global databases / seed vaults like Svalbard. Juxtaposing this hedge on future catastrophe to the one imagined in Ozeki's novel, I highlight the political possibilities for recasting "data" as "narrative."

Chapter 4 turns to the question of plastic, that wonder-turned-demon of recent years. Paradoxically ubiquitous and out of sight, seemingly benign and ambiguously toxic, plastic offers an iconic and illusive object for the representation of risk. This chapter is framed by reference to the Pacific Garbage Patch, a floating "continent" of garbage that, despite its size, offers a challenge to representation, not only because of its relative remoteness but also because of the manner in which the plastic appears, below the surface, in tiny particles, its risks at once visible (in, say, the plastic-filled stomachs of birds) and elusive (in the attractions of the particles and other toxins or in the potential chemical effects of the plastics themselves). Grappling with the representational complexity of the plastic problem, this chapter turns to the strategy of anthropomorphism, the depiction of this ultimately "morph-able" product as imbued with quasi-human characteristics. I focus particularly on the sympathetic plastic "characters" in a number of texts—including Ramin Bahrani's short film *Plastic Bag* (2009); Rachel Hope Allison's graphic novel *I Am Not a Plastic Bag* (2012); Karen Tei Yamashita's novel *Through the Arc of the Rain Forest* (1990); and Adam Dickinson's poem "Hail" (published in *The Polymers*, 2013). These contemporary iterations of the object narrative offer instances in which, as Gay Hawkins puts it, "Plastic bags [and other plastic objects] have their say," quite literally.[81] Here, drawing on the implicitly Althusserian tones of Adam Dickinson's poem, I ask what kind of environmental subject is interpellated by these animate plastic objects.

The final chapter builds on the *Bulletin of the Atomic Scientists*'s bookending of first and second nuclear ages by exploring the ambivalent relationship of climate change to the nuclear. Here, I turn to some of the troubling cultural production that accompanied the pre-Fukushima "nuclear renaissance"—from the industry's own "clean air energy" advertising campaigns to endorsements of nuclear energy by such environmental advocates as James Lovelock, Stewart Brand, and Patrick Moore. Pointing to the amnesia that accompanies these enthusiasms, I return to the origins of the first nuclear age, this time examining narratives of uranium mining in Canada, the source both for the atomic weapons used in World War II and for some of the material that fueled the Fukushima reactor. Tracing three narratives of the history of uranium mining at Port Radium in the Northwest Territories, Peter Blow's documentary film *Village of Widows* (1999), Marie Clements's play *Burning Vision* (2003), and David Henningson's subsequent documentary *Somba-Ke: The Money Place* (2006), I suggest, with Peter van Wyck, that this site might indeed provide something of a paradigm case for thinking risk today.

In the spirit of the reflexive, recursive nature of the project as a whole, the afterword is focused, not only on what comes "after," but also on what might have come before. I turn therefore to what I read as a sympathetic precursor to the project, the work of Japanese husband-and-wife collaborative artists Iri and Toshi Maruki. Best known for their series of murals depicting the aftermath of the atomic bombs in Hiroshima and Nagasaki, the Marukis went on to do a number of subsequent murals, from the Holocaust to the Rape of Nanking to the effects of Minamata disease, all of which they reportedly saw as part of a larger project of "painting the bomb." Reading this as an anticipation of the *Bulletin*'s "second nuclear age," I use the Marukis' art—which I then read through its ekphrastic representation in the work of contemporary American poet Ronald Wallace—as a figure for my own practice. In the process of these readings, risk emerges, not just as a means to assess hazard or benefit, but as a category through which to think more properly literary concerns—like irony, metaphor, metonymy, anthropomorphism, and analogy—even as it reminds us of the impossibility of separating these tropes from their historical contexts and political implications.

These excavations of the tropological wagers of a world at risk are themselves, of course, wagers, bets on the future of the archive. And as a precaution I should say that my focus on the dangers of risk should not be seen as a rejection of risk per se. I am not suggesting that there ought to be no risk, or that precaution could produce a world fully controlled and predictable, safe from calamity or harm. Precaution is, after all, less about controlling the future than it is about acknowledging and taking seriously the fact that there is no such control. Risk criticism itself is a risky practice; part of my wager is that some risks are worth running. In some sense precaution attempts an inversion of Derrida's "reflex and reflection," asking whether

1. Doomsday Clock. (Courtesy of the *Bulletin of the Atomic Scientists*.)

we might reflect in advance, taking seriously the uncertainties of the future, both by attending to the revelations of the past and by imagining the revelations that will have been to come. As Beck suggests, in grappling with risk society, "To knowledge drawn from experience and science we must add imagination, suspicion, fiction, and fear"[82] — and, one might add, hope.

ONE | **The Second Nuclear Age and Its Wagers: Archival Reflexions**

And as wager [gageure]. The archive has always been a pledge, and like every pledge [gage], a token of the future.
—Jacques Derrida, *Archive Fever*[1]

It would be a delight to know the future.
—J. Robert Oppenheimer, "Prospects in the Arts and Sciences"[2]

RISK'S IRONIC ARCHIVE

At the end of his magisterial novel of the Cold War, *Underworld* (1997), Don DeLillo imagines, in the sort of miraculous wish-fulfillment only possible in fiction, a post–Cold War scheme in Russia to destroy nuclear waste by exploding it with nuclear weapons: "The fusion of two streams of history, weapons and waste," explains Viktor, the entrepreneurial tour guide: "We destroy contaminated nuclear waste by means of nuclear explosions."[3] The nuclear menace is here turned in on itself, as one megarisk of the twentieth century annihilates another, closing out the nuclear age and the Cold War in one fabulous fell swoop. If only it were so. Though the Cold War may now be history, that the atomic age is far from over is likely clear to anyone reading the news on a daily basis. From former president George W. Bush's interest in nuclear "bunker busters" and warning that the "smoking gun" for Saddam Hussein's WMDs would likely be a "mushroom cloud," to President Barack Obama's negotiations with Russia regarding the nuclear arsenals of the former Cold War adversaries, to risks of new nuclear club members anticipated and realized in Iran and North Korea, nuclear weapons are still very much on the international agenda. And, even following Fukushima, so too is nuclear power generation, rediscovered and recast as a "clean and green" alternative to fossil fuels that even cofounder of Greenpeace Patrick Moore can support, as nuclear waste is transformed into a pesky technical challenge to be saved for a later day. But nuclear waste is one of the few things that even a nuclear explosion would not annihilate, and from the debates raging in my own small community in Canada over uranium mining, environmental hazard, and native land rights, to the medical reports emerging from the battlefields of Iraq of the pernicious health effects of depleted uranium dust; from controversy regarding waste storage at Yucca Mountain to the discovery of larger-than-expected stores of deadly plutonium beneath the former weapons-manufacturing facility at Hanford, Washington; from the radiation-detecting badges worn by Fukushima children to the "hot" tuna discovered off the coast of California, the persistence of the nuclear is clear.

How might we, literary and cultural critics, understand nuclear risk today—especially as it is cast so variously, as anachronistic stockpile, mobile new weapon, toxic pollutant, and "carbon-neutral" environmental salvation? And particularly as it is joined, and even partially subsumed, by fear of other megarisks? If, as DeLillo suggests, nuclear weapons and waste are two "streams of history" associated with the Cold War, what sort of freshets might accompany the newer incarnations of these twin demons? Beginning here, at the start of the *Bulletin*'s "second nuclear age," I will be looking backward, reviewing the first nuclear age from the vantage of a risk society in which the nuclear will have been, perhaps, one among many apocalyptic possibilities. The synecdochal nuclear is, in this context, at once illuminating and potentially obfuscating, for while the nuclear does offer a telling model for other global risks—it was, after all, the technology that inaugurated the secular promise of human extinction—it also signifies differently when it moves from whole to part. Climate change and other megarisks (biotechnology, nanotechnology, the ozone hole) are not simply analogous, for these risks, emerging in peace, accidentally, as unintended side effects of other intentional processes, comprise a different archive. If we may once have believed that the end of days would come in a blaze of nuclear firestorm (or the chill of the subsequent nuclear winter), we now suspect that the apocalypse may be much slower, creeping in as chemical toxin, climate change, or bio- or nanotechnologies run amok, and in this new company, it is perhaps not the fiery blast but the radioactive spill that moves to the fore. In an age of risk, the apocalypse cannot be avoided through the "rational" deterrence of mutually assured destruction, for the destruction in this case is unintentional, coming as a consequence less of the dramatic decisions of the strategists than of the mundane results of peaceful business as usual—the push of an aerosol spray, the flick of a light switch, the disposal of a laptop computer.

Confronting such an enlarged archive is a daunting task: Are we, mere literary critics, modest humanists, now responsible, not only for accounting for the follies of the nuclear—from "duck and cover" to the technical differences between fission and fusion, and among atomic, hydrogen, and neutron bombs; from the minutiae of START talks to the speculative fictions of MAD—but now, too, for the significance of CO_2 levels, the operations of endocrine disruptors, the proliferations of GMOs (with all of their Frankensteinian cultural freight)—not to mention the implications of the new synthetic biologies pioneered by Craig Venter of gene-mapping fame? My purpose here is not to be exhaustive, to catalog all risks that might fall—or befall us—in this capacious second nuclear age. Instead, I intend to map a few gatherings—of texts, historical moments, technological advances and missteps—in the interest of querying anew the contours of a first nuclear age now reconceived in the light of a second. In the spirit of Beck's "second modernity"—a moment "reflexive," both reflecting upon itself and confronted by its own side effects—I will consider the second nuclear age, in part as a successor to the first, perhaps, but also as a reflexive return, for even as the addition of the newer risks inevitably affects what will be consigned (in Derrida's sense of *"con*signing" or *"gathering together signs"*)[4] to the archive in the second nuclear age, it may also affect the future of that earlier archive, transforming what the first nuclear age will have been as well.

That global risk might comprise an archive for which literary critics are responsible at all is, of course, counterintuitive. As we know, the figure likely to have the kind of specialized expertise to assess the dangers of risk society is not the humanist—or even the social scientist—but the "hard" scientist, the one whose specialized knowledge of ice cores or rems puts him or her in the position of revealing the hidden truth behind aberrant temperatures or apparent cancer clusters. Given its invisibility and the specialized equipment necessary for its detection, risk forms, as Ulrich Beck argues, a kind of "'shadow kingdom,' comparable to the realm of the gods and demons in antiquity, which is hidden behind the visible world and threatens human life on this Earth."[5] In this context, the scientist is not one figure among many, for he or she is the one who has access to this invisible but no less real world. This puts the scientist in risk society in roughly the position that Michel Foucault described—appropriately enough in terms of the atomic scientist—as the bridge between the "specific" and the "general" intellectual: "It's because he had a direct and localised relation to scientific knowledge and institutions that the atomic scientist could make his intervention; but, since the nuclear threat affected the whole human race and the fate of the world, his discourse could at the same time be the discourse of the universal."[6] The discourse on the universal is here grounded in that knowledge derived from the truths of science, which leaves the nonscientist, the risk critic, in the awkward position of intervening only at a later stage, *"dependent on second-hand non-experience controlled by professionals outside their field,* with all the damage that does to their battered ideals of professional autonomy."[7]

But the scientist of first modernity—or, in this case, the first nuclear age—is not the scientist of the second, for the sciences, like modernity itself, have become, Beck argues, reflexive: "When they go into practice, the sciences are now being confronted with their own objectivized past and present—with themselves as product and producer of reality and of problems which they are to analyze and overcome"—thus the miracle coolants and plasticizers of yesteryear produce the nightmarish ozone holes and endocrine disruptors of today (and perhaps today's GMOs and hand sanitizers tomorrow's biopollution and superbugs).[8] As "everywhere, pollutants and toxins laugh and play their tricks like devils," the scientist is at once paradoxically both the exorcist who might save us and the conjurer who set these demons to work—a figure who, even acting in the best faith and with the best of intentions, seems now unable to control the genie let out of the bottle or predict its actions or consequences.[9] The risk critic thus must balance a stance at once *"critical* and *credulous* of science,"[10] as we must believe in the veracity of this "second-hand non-experience" and, cognizant of past precautionary failures, doubt it at the same time.

As foreign as the chemical formulas and statistical charts, the ice cores and rems, might be, then, there is something in risk that may in fact be appropriate to our particular skill set. If, as Hayden White once suggested, all historiography involves a "tropological wager," an implicit bet on the explanatory potential of one of Burke's master tropes,[11] this is certainly true in the case of Beck's risk society, in which trope and wagering are tightly connected. Commenting, in a 2006 lecture at the London School of Economics, Beck describes the narrative of risk in tantalizingly literary terms. Beck asserts: "The narrative of risk is a narrative of irony [that] deals with the involuntary satire, the optimistic futility, with which the highly developed institutions of modern society—science, state, business, and military—attempt to anticipate what cannot be anticipated."[12] As contemporary "post"industrial society habitually and inadvertently generates hazards it can neither imagine nor control, the discourses of risk management continue to proceed with business as usual. Members of world risk society—that is, all of us—are thus confronted with the juxtaposition of an ever-growing catalog of unanticipated disasters—from the wonders-turned-horrors of asbestos and CFCs to the hazards of nuclear and biotechnologies to mad cow disease, salmonella, and bisphenol-A—and an ever-renewable risk-optimism—particularly in the techno-fixes imagined to counter existing risks, from carbon sequestration to nanotechnology—an irony that would almost be comedic if it weren't so deadly serious.

Reading the narrative of risk thus involves attentiveness to the situational ironies of the archive, an interpretive practice in which humanists may, indeed, have some expertise. Irony is, of course, hardly exclusive to theories of risk. Beck's model resembles D. C. Muecke's characterization of the "generalized irony" of events, in which, because the "future is essentially unknowable or inescapable," "there can be no confident anticipations."[13] Such, for Muecke, is the nature of the human condition in general, regardless of historical circumstance. Life is a wager, and even the most cautious of forecasts cannot forestall the "being struck by a meteorite" of unanticipated risk.[14] But for Beck, the vagaries of the future are not those of blind fate or meteorites. They are, rather, the side effects of modernity come ironically back as a menacing environment with which we are now confronted. The problem is thus not that we are presently faced with some newly ironic human condition. Ironies of situation of the sort that characterize risk society are still analogous to those that plagued Oedipus. The problem is that those "highly developed institutions" to which Beck refers have bet against irony in a very high-stakes wager on the knowability of the future.

The nuclear archive is full of ironies, whether verbal, situational, or dramatic: apocalyptic futures seem to inspire this mode of engagement.[15] But the nuclear age that persists beyond the Cold War, the moment in which the nuclear is joined by risks that crept in alongside it, gradually escalating to reach its grand scale, might remind us of the power of the revelations of situational irony to change, not just the present or future, but also the past. Turning, here, to that iconic general intellectual, the atomic scientist, I will argue that a return to the early moments of the first nuclear age provides a useful origin for the archive of global risk, not only as a template for the structure of subsequent hazards, but also for the potentially dialectical possibilities for an alternative future that those hazards might ironically offer. The atomic scientists' own archival additions are instructive, here, for they attempt to grapple with the questions of responsibility and futurity that preoccupy us still, but it is in reading the reflexive returns to these scientists—in fiction, anecdote, metaphor—that we, literary critics in the risk society, might find our own "specific" knowledge granting us "universal" currency, as these reflexes prompt further reflections on the relationships among diverse global risks and the "whole human race and the fate of the world." The heroic (or tragic) narrative of the atomic scientist on the one hand and "the world" on the other is a story we have inherited from the first nuclear age; my wager here is that revisiting it in the second, cognizant of the multiplication of risk and the unevenness of its distribution, might offer a way to begin to figure the contours of the illusive (and allusive) archive of the second nuclear age.

ORIGINS

In some ways, of course, the very concept of an archive of risk is ironic, given that, as Derrida points out in *Archive Fever*, the term "archive" is rooted in origins, its "arché" at once referring to "commencement" and "commandment," both a historical moment and an institution or expression of law. For Beck, however, risk society does not emerge suddenly and dramatically "in the manner predicted in the picture books of social theory"; rather, the transition from industrial society to risk, first modernity to second, occurs *"on the tiptoes of normality, via the back stairs of side effects."*[16] Risks are often unacknowledged until it is too late; catastrophes emerge unexpectedly, accompanied less by "commandments" than by public apologies, calls for international aid, or postmortem exposés. If the first nuclear age itself has at least the illusion of a definitive beginning—often cited as the first explosion of "the gadget" at Trinity in 1945, a commencement accompanied by J. Robert Oppenheimer's iconic commandment, "Now I am become Death, the Destroyer of worlds"[17]—the inaugural moment of risk is less clear.

Beck himself is cagey about where, precisely, second modernity might begin. But another theorist of risk, Sheila Jasanoff, has hazarded an originary moment that might provide both ironic commencement and ironic commandment for a world at risk. In an article appropriately titled "Risk in Hindsight," Jasanoff rereads the Trinity explosion in light of subsequent history, recasting it as the origin, not only of the first, but also of what we might, now, call the second nuclear age:

> It is hard to date the precise moment of emergence of incalculable risks, but a turning point may be the explosion of the first nuclear weapon at the Trinity test site in the New Mexico desert on 16 June 1945. That initial public demonstration of the results of the Manhat-

tan Project alerted the world to the possibility of total annihilation. It is said that, moments before the blast, the eminent physicist Enrico Fermi "began offering anyone listening a wager on 'whether or not the bomb would ignite the atmosphere, and if so, whether it would merely destroy New Mexico or destroy the world.'" Since then, not only the continued threat of a nuclear holocaust but a succession of more or less devastating natural and human-made disasters have kept alive the spectre of essentially incalculable, and hence uninsurable, risks.[18]

Jasanoff's story has a regressive quality. The site of the origin of world risk is Trinity, and at first the key moment seems to be the blast itself, the demonstration ("public" only for the immediate scientific community) of the terrifying power unleashed. But though that moment might be a logical one for thinking the origins of the atomic age, it is not the appropriate commencement for incalculable global risk. Instead, winding the clock back to "moments before the blast," Jasanoff returns us to a moment of nonknowledge, before the revelation of the weapon's spectacular power.

Fermi, a thoughtful man, winner of the Nobel Prize, famously the creator of the first "atomic pile" (an event that proved a reaction could be sustained), much loved by his students, likely did not intend for his wager to become the "commandment" accompanying the "commencement" of an archive. Notice the layering of citation in the quotation, the passive voice that obscures the informant ("it is said"). It was an aside, a joke, an offhand comment before a secret event. It comes down to us secondhand, a side of the bomb that was not to be told. The arché of Jasanoff's archive is thus itself a kind of accident, a leak, a spill. Intent is fuzzy in accounts of the wager, for Richard Rhodes, in his influential account, *The Making of the Atomic Bomb* (1986), suggests that the incalculability of the risk was "Fermi's point."[19] But even if his wager had serious intent, that risk was not an argument against exploding the gadget, an ambivalent relationship to the future that Muecke describes as basic to irony: "Not only can we *not* trust the future; we *must* trust it."[20] Risking futurity itself, Fermi's wager reveals the nature of the stakes, becoming, for Jasanoff, the specter that haunts risk society, uncannily both the past and the uncertain future.

In its original context, Fermi's wager is, of course, a ruse, an instance of clear verbal irony; there is only one way to bet, since if one were to bet on either of the apocalyptic results, one could never collect. But the joke, in retrospect, may, Jasanoff's reading suggests, be on Fermi. Read in the context of subsequent history, the situational ironies about which Fermi and the other atomic scientists could only have speculated seem to trump whatever dark humor might have been intended in his wager. In fact, if, in Muecke's ironic universe, we cannot trust the future, we also, it seems, cannot trust the past, not only because, given the unprecedented nature and scale of contemporary risks, past experience cannot provide a guide, but also because the past has made wagers that we now must pay out—not, perhaps, the igniting of the atmosphere, but the destruction of Hiroshima and Nagasaki, the pervasiveness of strontium 90, the quarantine of Chernobyl, or the expulsion of contaminated groundwater from the still volatile Fukushima plant.[21]

FIGURING FERMI'S WAGER

In Jasanoff's origin story, the template for risk society is not, then, precisely the bomb, but the willingness to wager on odds not only incalculable but inconceivable, the archive of which would surely include the bomb but also other risky technologies in war and in peace. And if this recasting of the origins of the atomic age as the origins of the age of risk has sociological and historical implications, it may also affect models of literary history, offering new insight on texts presumed to be solidly situated in the Cold War context. Indeed, in hindsight, a text like Kurt Vonnegut's *Cat's Cradle* (1963), which figures an atomic scientist as one of its primary preoccupations, might be read, not just as a critique of the nuclear but also as a model for thinking that technology in a context of global risk. Rereading this text in the second nuclear age illuminates the ways in which Vonnegut in fact figures an archive and an archival practice in the midst of a transition, as the protagonist confronts the ironies and absences in an atomic archive that subsequent technologies are in the process of transforming. Thus, *Cat's Cradle*, in its satirical critique of the wagers of risk society, presents a model for a reflexive response to the "involuntary satire" that Beck suggests characterizes the "narrative of risk."

Published just after the Cuban missile crisis in 1962 (the event often cited as bringing the United States and Soviet Union as close as they ever came to nuclear war), *Cat's Cradle* is often read, as William Deresiewicz has recently reminded us, as "an allegory about nuclear weapons—refigured here as *ice-nine*, the crystal seed that makes the waters freeze—and an indictment of scientists who evade responsibility for the consequences of their discoveries."[22] Indeed, writing in another hot moment of the Cold War, in his 1986 article on Vonnegut's novel, "Rescuing Science from Technocracy," Daniel Zins, himself a noted nuclear critic, employs a quote from Fermi as an epigraph: "Don't bother me about your conscientious scruples. After all the thing is beautiful physics."[23] The implication seems to be that Fermi is an appropriate real-life counterpart to Vonnegut's imagined atomic scientist, Felix Hoenikker, a man comically deaf to the ethical implications of his work. But if we take, not the Fermi without "conscientious scruples," but the Fermi of the ironic wager, then "Fermi" might be said to animate not just the target of Vonnegut's satire but the satire itself. If, following Rhodes, incalculability was Fermi's point, it certainly seems to be Vonnegut's point as well. Refracting this incalculability, as Jasanoff does, in hindsight, as characteristic of a more generalized risk helps to explain the otherwise puzzling gap between the vehicle of ice-nine and the tenor of the bomb, now viewed less as replacements for each other than as a series that other risks might join, the novel now resituated in the archive of the second nuclear age, speaking as clearly to the proliferating catastrophes of risk society as it may have to the particularities of Cold War nuclear fears.

Cat's Cradle begins with a pretense of archiving the atomic age. The protagonist, John, reports that he had set out to write a book titled *The Day the World Ended* about "what important Americans had done on the day when the first atomic bomb was dropped on Hiroshima."[24] That original book, he explains in a telling use of the past progressive, "was to be factual," but as with so many predictions and plans, this one went awry, and he has chosen instead to write the book we are now reading, presumably and cryptically titled *Cat's Cradle* instead. As it turns out, "the day the world ended" would have been an appropriate title for this latter book as well, for what John discovers in his research on one of the "fathers" of the atomic bomb, the fictional Felix Hoenikker, is that this scientist has conceived a second, even more terrible technology called ice-nine, a seed crystal that alters the melting point of any water with which it comes into contact to 114.4 degrees Fahrenheit. Daniel Zins's epigraph is here highly appropriate. Vonnegut's Hoenikker is indeed untroubled by scruples, and there is no doubt that Vonnegut's satire targets what he elsewhere calls "morally innocent scientists," those who could work on the technical problems associated with weapons systems without considering the ethical principles they might thereby be violating. Referring, in a speech delivered a few years after the novel's publication, to Louis Fieser, the inventor of napalm, Vonnegut suggested that there was "nothing at all sinful in Dr. Fieser's creation of napalm," simply because "scientists like him were and are as innocent as Adam and Eve."[25]

But ice-nine is not napalm. Though it may have been originally a military technology, it is not intended to be a weapon. Rather, as Asa Breed, Hoenikker's former colleague, explains, it was commissioned by a general in the Marines: "The Marines, after almost two-hundred years of wallowing in mud, were sick of it. . . . The general, as their spokesman, felt that one of the aspects of progress should be that the Marines no longer had to fight in mud."[26] Of Dr. Hoenikker's response, Dr. Breed comments, "In his playful way, and all his ways were playful, Felix suggested that there might be a single grain of something—even a microscopic grain—that could make infinite expanses of muck, marsh, swamp, creeks, ponds, quicksand, and mire as solid as this desk."[27] Not surprisingly, this technology not only kills its maker, but escapes the inept grasps of his human progeny, eventually freezing all water and most plant and animal life on the planet, leaving only a small nonreproductive human population to witness the end of life on earth. The characters' Promethean aspirations—political, technoscientific, religious—all come to a cataclysmic end in a deep freeze. The novel's premise is thus that it "was to be" a book of the atomic age, firmly rooted in the history of August 6, 1945, but it has turned, instead, into an odd allegory, an extension, an elaboration, at once fictional and extrapolative and uncannily familiar.

Cat's Cradle is clearly a product of a particularly hot moment in the Cold War, and certainly the cataclysmic results of ice-nine are influenced by the apocalyptic zeitgeist of the period. It is thus tempting to read *Cat's Cradle* as a text solely of the atomic age, with the ecological impact of the doomsday technology itself a kind of side effect of the real concern with nuclear war. But that nuclear weapons were not the only imminent threat was increasingly apparent in the early 1960s as well—1962 was also the year in which Rachel Carson's exposé on the dangers of herbicides and pesticides was published serially in the *New Yorker*. Vonnegut's choice to figure, not the bomb, but an alternative technology, a technology that emerged less as a weapon than as a time- and energy-saving device, suggests that his concerns may be as much with those risks that, as Beck puts it, "enter the world peacefully"[28] as with those that enter the world through war—or, as in the case of Carson's chemicals, that move from military to civilian use. Ironically enough, of course, the problem with which we are now faced is one even Vonnegut could not forecast, not a horrible deep freeze (itself no doubt a play on the "locking" of the Cold War), but an unprecedented melt, both of international relations and of polar ice packs—a "liquid modernity" quite literally.[29] But this is arguably the point of the novel: no calculation can forestall the ironies of risk. A commemorative plaque on the wall of Hoenikker's old laboratory ironically sums up the risk represented by the "morally innocent" scientist: "The importance of this one man in the history of mankind is incalculable."[30]

The incalculability here is, clearly, ironic, an intervention in the heroic narrative of the influential scientist. And Vonnegut's novel might be read, then, as an experiment, not only in representing global risk, but in representing its archiving in the face of an open future that perpetually transforms the past. The archival narrowing with which Vonnegut's John begins—on "what important Americans had done on the day when the first atomic bomb was dropped"[31]—turns out to be shortsighted, not only in its focus on the bomb, but in its preference for "important Americans." If, for Foucault, the atomic scientist navigated a position between his specialized expertise and the well-being of the "whole human race and the fate of the world," Vonnegut highlights the way such responsibilities might go awry, but the novel also suggests problems with the abstract generalization of "the world," a place that, even while it is, in narratives of total thermonuclear war, "one," is also, in the nonlocalizable nuclear war on the ground—and its risk corollaries—variably and unevenly at risk.

One might, then, read the opening of *Cat's Cradle* as satirizing what Ken Cooper has called the "compartmentalized and essentially Olympian story of Los Alamos," a version of American history that cordons the nuclear off from other historical archives. Cooper challenges this Olympian story by interfacing that archive with the one devoted to civil rights, pointing, in the process, to the fact that the seeming "whiteness of the bomb" is possible only if one ignores the labor that supported the community at Los Alamos. Reading across a number of atomic age literary texts—from Langston Hughes's "Simple" stories (published in the 1960s), to Ishmael Reed's *Mumbo Jumbo* (1972), Paule Marshall's *The Chosen Place, the Timeless People* (1969), and Leslie Marmon Silko's *Ceremony* (1977)—Cooper supplements the atomic archive, both by revealing the ways in which the history of the bomb is inseparable from the history of oppression and discrimination and, following these writers, by tracing what the bomb represents backward in time, undoing the fetishizing of Trinity as origin, the "zero hour" of a new age.[32]

Vonnegut's choice to set the latter part of *Cat's Cradle* in the fictional impoverished Caribbean island of San Lorenzo, emphasizing its neocolonial relationship to the United States, begins to suggest this wider view, as histories of U.S. imperialism accompany present export of environmental hazard, but the cartoonish depiction of the island and its people perhaps undermines what might be a more serious critique. Nevertheless, if Cooper points to the presumed "whiteness of the bomb," the reader of *Cat's Cradle* might point to the "whiteness of ice-nine"—not only does it turn all that it touches an icy hue, but it seems, in the end, to privilege whiteness, or at least to suggest that the white American characters are the only people willing to live under the utterly unsustainable ecological conditions of a frozen planet.

As a critique of the constitutive absences of the atomic archive, though, Vonnegut's novel might usefully be supplemented by texts more attentive to the unevenness of global risk. Countering the innocence of a Felix Hoenikker, that atomic scientist who need feel no responsibility for the practical consequences of his research, we might draw the more cynical figures from a text like Toni Cade Bambara's *The Salt Eaters* (1980), a novel that Cooper cites in passing. *The Salt Eaters* offers a clear environmental justice response to the idea that issues of nuclear and environmental risk are somehow separate from the politics of race. Exasperated with the talk of nuclear apocalypse, Bambara's imagined activist Ruby says: "All this doomsday mushroom-cloud end-of-planet numbah is past my brain. Just give me the good ole-fashioned honky-nigger shit. I think all this ecology stuff is a diversion."[33] Her friend Jan replies, clearly in the spirit of the novel itself: "They're connected. Whose community do you think they ship radioactive waste through, or dig up waste burial grounds near? Who do you think they hire for the dangerous dirty work at those plants? What parts of the world do they test-blast in? And all them illegal uranium mines dug up on Navajo turf—the crops dying, the sheep dying, the horses, water, cancer."[34] Her queries are, of course, rhetorical, informing both Ruby and the reader that the "diversion" is in the idea that social and ecological risks are separable.

Bambara's more pointed critique of environmental injustice extends to her version of those responsible for contamination. Whereas Vonnegut's Hoenikker plays "puddly games" with ice-nine, innocently informing those present at his Nobel Prize speech that he "never stopped dawdling like an eight-year-old on a spring morning,"[35] Bambara's nuclear engineers play a different "repertoire of games" that are "peculiar to their profession"—no less dangerous but potentially less innocent.[36] Their games, focused not on inventing new technologies but on manipulating the effects of the technology that is extant, include "Fail-Safe Phooey," "Fission," and "Fix," each a drinking game that imagines scenarios that the players have to work their way through—or pay for the round. "Fission" is fairly self-explanatory—the players imagine ways of producing chain reactions with fissionable materials—but "Fail-Safe Phooey" and "Fix" are more cynical, the former involving overriding fail-safe systems, and the latter dealing with the aftermath of an accident, either by actually "fixing" it, by "dismantling the plant and disposing of the contaminated parts," or, more likely, putting in a "fix," "that is, hire a team of experts to conduct a study proving that the defective parts were neither vital nor even necessary to plant operations." "Fix cards," we learn, "could be purchased or traded."[37] The implication in *The Salt Eaters* is, of course, that such games are played as a kind of practice for the real thing—in fact, that the practice and the real are essentially the same thing. In this, they share a cavalier willingness to wager the future with Vonnegut's playful Hoenikker, who asks, "Why should I bother with made-up games when there are so many real ones going on?"[38] Though I take Cooper's point that the scientist as origin risks replicating the heroic/tragic narrative of the atomic age, then, I would argue that these novels usefully focus our attention on that figure, not to replicate the Olympian story, but for the purpose of ironic critique.

This model of the scientist—the version Vonnegut's Asa Breed accuses the narrator, John, of caricaturing: "heartless, conscienceless, narrow boobies, indifferent to the fate of the rest of the human race, or maybe not really members of the human race at all"[39]—is, of course, an oversimplification. Nevertheless, the "pure" and "innocent" scientist, driven by "beautiful physics" and untroubled by "conscientious scruples," persisted, Bambara's more cynical figures notwithstanding, as an alibi on into the atomic age. Writing, seemingly without ironic tone or biting satirical impulse, in his antinuclear opus *The Fate of the Earth* (1982), Jonathan Schell offered a similar model for innocence in scientific innovation: "Scientific findings, some lending themselves to evil, some to good, and some to both, simply pour forth from the laboratories in senseless profusion, offering the world now a neutron bomb, now bacteria that devour oil, now a vaccine to prevent polio, now a cloned frog."[40] Schell concludes: "Although it is unquestionably the scientists who have led us to the edge of the nuclear abyss, we would be mistaken if we either held them chiefly responsible for our plight or looked to them, particularly, for a solution."[41] Scientists, here, are indeed the Adam and Eve of modernity, innocent instigators of the "senseless profusion" with which we, laypeople, are left to grapple.[42]

Jasanoff's return to the moments before the Trinity explosion shifts the emphasis of the atomic age from the bomb to Fermi's wager—a wager, in this case, against wagering, a verbal irony that worked to foreclose ironies of situation—which, in retrospect, is about more than the bomb. It is a tone-deafness to consequence, and whether this is fair to Enrico Fermi, it is undoubtedly descriptive of Vonnegut's comically flat Hoenikker. But though Vonnegut's scientist characters—and their presumed nonfictional counterparts—are clearly pilloried in the novel, there is a sly reference to one very real atomic scientist whose moral response to the bomb is fairly well known. At one point, Hoenikker's son, Newt, is recounting anecdotes of the atomic age for John, the writer-protagonist. Lamenting that John is interested only in the events of August 6, 1945, Newt recalls a story from an earlier moment, reportedly just following the Trinity explosion: "After the thing went off, after it was a sure thing that America could wipe out a city with just one bomb, a scientist turned to Father and said, 'Science has now known sin.' And do you know what Father said? He said, 'What is sin?'"[43] Though the punch line of the anecdote is Hoenikker's amoral response, the story draws on the actual moral musings of the scientist who became notorious for mixing his "conscientious scruples" with "beautiful physics," J. Robert Oppenheimer.

THE DIALECTICS OF IRONY: "ONE WORLD OR NONE"

A turn to Oppenheimer offers a somewhat different model of the wagers of irony, and one that uncannily presages, again, the contemporary context of risk society. Though the first step, for critics like Ulrich Beck, is a gathering together, an archiving of isomorphic examples, founded, as in Jasanoff's model, on the generally losing wagers against irony, the second step is what Beck, in his own wager, believes might be an ironic twist in the plot of the narrative of global risk.[44] Faced with the high stakes of the wagers being made, we have, Beck suggests, three options: "denial, apathy, or transformation."[45] Optimistically, Beck chooses the latter, positing, "against the grain of the current widespread feeling of doom," that there might, dialectically, be an "enlightenment function" to global risk, whereby the very extremity of the danger leads to an alternative, a cosmopolitan "subpolitics" outside of the confines of particular nation-states, a global recognition of being-at-risk that might lead to a global transformation. A turn to the archives suggests that this dialectical irony might also find a correlate in the first atomic age, for if the early atomic age can provide a kind of template for incalculable risk, it might also provide an instance, whether cautionary tale or instructive model, of an attempt to imagine a cosmopolitan alternative, and one dialectically—and ironically—supposed to emerge from just that risk. In the immediate aftermath of World War II, as the world struggled to come to terms with the meaning of the new and terrible weapon, some of those who produced it, who knew its awesome and terrible power, drew together to form what was called the "scientists' movement," a political impulse that drove scientists from the secret labs of the Manhattan Project out into the public sphere—the popular media, the Congress, the United Nations—to advocate disarmament. These scientists were determined that, as rational men who had produced the technoscientific marvel, they might also have a hand in interpreting its meaning and reengineering a society able to cope with its consequences.[46]

If nuclear war became the "unthinkable," it was in part due to the rhetoric of these atomic scientists who argued that, given the likely future ease and cheapness with which nuclear weapons could be manufactured, and given the quickly escalating size of the potential explosions, nuclear weapons offered the world an opportunity to overcome the sort of conflicts that had led to the two world wars previous. Given the totality of the wars to come, the scientists argued, humankind was faced with two choices: the creation of a world government or an arms race that could only end in total annihilation. Fear of that potential global apocalypse would bring humanity together under a common purpose—the newly formed United Nations seemed to be a logical start—or apocalypse was nigh.

That there was something fairly messianic about the scientists' movement is clear in D. R. Davies's 1947 introduction to the anthology *One World or None*, a phrase that became shorthand for the scientists' collective position: "It is the scientists, the brass-tacks men, who are talking in New Testament language about the end of the world. . . . When scientists abandon their cloistered laboratories to shout in the market-place, then there must be something very grave and unusual afoot—and we would be well-advised to heed what they are saying."[47] As hubristic as this messianism might be, though, there is something at least embryonically reflexive here, as the scientists confronted the new environment that their innovation had produced. Gathered up in the bomb—virtually engineered into it—was a deep ambivalence, a potentially deadly pharmakon, poison or cure for the problems of national sovereignty that had led to wars in the past. "We have made a thing," Oppenheimer told those gathered at a symposium on atomic energy in November 1945, "that by all the standards of the world we grew up in is an evil thing," a thing in which the "elements of surprise and of terror are as intrinsic to it as are the fissionable nuclei."[48] But Oppenheimer also brought good news, for this evil was also going to be "our great hope" for a world without war, a world in which war must, necessarily, be obsolete.[49] The bomb came to represent, for Oppenheimer and others involved in the scientists' movement, this "peril and promise" of total war or perpetual peace.[50] According to Charles Thorpe, such logic drove the invention of the bomb as much as its subsequent rationalization: "If violent antagonisms were rooted in 'history and traditions' then science, transcending these ancient divisions, offered hope, the only hope, for peace.... The bigger the bomb, the greater the rupture with history, the more hopeful the new dawn."[51] In this spirit, as Eugene Rabinowitch put it, the "founding of the *Bulletin of the Atomic Scientists* was a part of the conspiracy to preserve our civilization by scaring men into rationality."[52]

Reading the documents of the scientists' movement in hindsight offers, of course, a different narrative of irony. The large and numerous weapons did bring peace, not the rational benevolent world state the scientists imagined, but the "hot peace" of the irrational MADness of deterrence and the wars by proxy in Korea, Vietnam, and the Middle East. The ironies of the scientists' movement were myriad: the bomb was to bring peace, as a common threat of human extinction would bring a world government capable of preventing it, but instead, as Rabinowitch reported just five years after Trinity, "While trying to frighten men into rationality, scientists have frightened many into abject fear or blind hatred."[53] Within a few years, the Soviet Union also had the bomb, the Cold War was under way, and Oppenheimer would be stripped of his security clearance, a moment that Daniel Bell suggested signaled that "the messianic role of the scientists ... was finished."[54] And though Oppenheimer cannot perhaps be held responsible for this future—a future that the fears the scientists' movement inspired nonetheless fueled—he can arguably be held responsible for his own unstated encouragement of "evil," for he was the chief signatory of the recommendation to use the bombs in Japan, countering the Franck report, which had advocated for Trinity as a truly public demonstration of the weapon's might. Thus, though Oppenheimer's story—from powerful "father" of the bomb, to responsible public intellectual and eventual opponent of the hydrogen bomb, to victim of the Cold War's red scare, stripped of security clearance and political influence—is often written as tragedy, there was undoubtedly something farcical in it as well. Whatever hope there might have been for a global antinuclear cosmopolitanism, a politics based on the fears produced by global risks, it quickly faded, and, as though delivering on the scientists' prophecy, the superpowers developed truly apocalyptic arsenals.

THE SCIENTISTS' MOVEMENT IN HINDSIGHT

Oppenheimer's ironic "peril and hope" was, perhaps, more messianic than reflexive, but, in the spirit of Jasanoff's reanimation of Fermi, now a figure for global risk, we might ask what the subsequent scientists' movement, their wager on the hope of the atomic age, might look like, recast in the hindsight of world risk society, a world in which peril is often, still, tied to hope. Such speculative imagining is, in fact, arguably the project of a more recent novel, Lydia Millet's *Oh Pure and Radiant Heart* (2005), a text that takes "reanimation" and "reflexivity" quite literally. An addition to the atomic archive that necessarily transforms it, *Oh Pure and Radiant Heart* situates the scientists' movement in an age of risk, in which, as in Vonnegut's and Bambara's novels, catastrophes have multiplied, moving from projected future to present reality. Millet imagines, in effect, a confrontation of the first and second nuclear ages, as three atomic scientists, Fermi, Oppenheimer, and Leo Szilard (the last of whom initiated the Manhattan Project with a letter, cowritten with Einstein, to FDR warning of the dangers of a Nazi atomic bomb) are transported from the site of the Trinity explosion to nearby locations in 2003, where they attempt to come to terms with the world their "gadget" has wrought. Millet's scientists do not know why they have been brought to 2003, but they speculate that there might be "something in the event, something anomalous and unprecedented" that made their time travel possible, a beginning of something that may—or, as the novel proceeds, may not—be ending.[55] A satire that targets both "them" and "us," both the scientists and the world into which they are summarily thrown, the novel bookends the first and second atomic ages with a kind of ironic messianic resurrection, as the "trinity" of scientists is brought to witness the slow apocalypse of the present. And for the layperson confronting the ironic paradox of brilliant, in some cases pacifist, scientists responsible for producing what perhaps remains the most dangerous technology of our times, *Oh Pure and Radiant Heart* offers an opportunity to imagine asking them that nagging question: "What were you *thinking*?"

Oh Pure and Radiant Heart offers a meta-archival exploration of the archive, drawing figures from the first nuclear age and reanimating them for the future toward which their own practices of archivization—from the creation of the bombs to the scientific, political, philosophical, and even, in Szilard's case, fictional documents that accompanied them—were invariably oriented. What Millet's imagined scientists discover is the history that, having been transported directly from Trinity, they could not themselves have witnessed, and that, as a consequence, they must read accounts of, which as Millet explains in an ironic reversal, they "read as though they were fiction."[56] The novel is full of historical detail, from statistics about the arsenals of the Cold War adversaries to excerpts from memoirs of Hiroshima survivors to headlines from major newspapers. And fittingly, Millet's protagonist, Ann, is a librarian whose interest in the period and access to its archive provide the device by which these historical details can be included. *Oh Pure and Radiant Heart* thus enacts something like a double defamiliarization: not only are the scientists in the position of seeing our world anew—and thereby allowing us to do so as well—but they are also in the position of seeing themselves—their post-Trinity selves—as others see them, as texts, additions to an archive to which they contributed but which they themselves never experienced. And in their return Millet imagines something like a return of history, as these scientists, learning of the existence of the massive arsenals that they themselves predicted might come to pass, attempt to launch an ultimately unsuccessful peace movement, a kind of farcical repetition of the impulses that led to the scientists' movement as well as to the formation of the *Bulletin of the Atomic Scientists* and its iconic Doomsday Clock.

If, as Derrida argues, "a spectral messianicity is at work in the concept of the archive and ties it, like religion, like history, like science itself, to a very singular experience of the promise,"[57] Millet's fictional archive and the scientists who emerge from it at once evince and ironize this promise. Whether the scientists have returned to do some penance for their own sins or to rescue us from ours—or both—is never fully clear, but in their return is something distinctly messianic, at least for those around them. Encountering this trinity of scientists, Ann (accompanied by her long-suffering husband, Ben) becomes a kind of guide and devotee, dispensing with her job and much of her savings in order to follow the scientists on their pilgrimage from Hiroshima, to the Nevada test site, to Washington, DC, on a new campaign for nuclear disarmament. Along the way, they gather a rather motley following of trust-fund hippies, holdover peaceniks, and Christian fundamentalists, the last of whom decide that the arrival of these scientists is a harbinger of the coming Rapture, which, naturally, they would like to usher in. And as what was initially to be a peace campaign gets increasingly diverse and populous, the movement falls apart amid gunfire and chaos. By writing the scientists into the plot of the apocalypse, Millet implicates them and their gadget in a seemingly inevitable end-times scenario, as their original sin at Hiroshima cannot be undone, even by marching on Washington.

Cast as characters in a novel and unmoored from their professional identities, Millet's scientists appear as collections of quirky character traits. Fermi is enigmatic, preferring to go on long hikes that remind him of the Italian countryside of his youth; he eventually withdraws entirely to watch birds at a secluded sanitarium. Oppenheimer, with his iconic porkpie hat and newly offensive cigarettes, emerges as the primary confidant for the protagonist, Ann; solid, reflective, and basically personable, his periodic philosophical insights mark him as Foucault's "rhapsodist of the eternal."[58] But the figure most comically over the top is the brash Szilard, a man of tremendous energy, hubris, and appetites, who consumes youth culture and donuts with equal relish. If Oppenheimer becomes the charismatic face of the renewed scientists' movement, it is Szilard who provides its energy, as he works behind the scenes and often in the bathtub to orchestrate events.

Such characterizations—and the tantalizing references to actual historical documents and personages of the atomic age—might inspire the critic to assume the role of Millet's protagonist, rummaging through the archive for clues to the interpretation both of the novel and of risk society in general. In bringing these characters together in a single, unified movement, Millet's novel bridges what Margot Norris calls the "fissured story" of the Manhattan Project, as the scientists at Los Alamos (Oppenheimer chief among them) and the scientists in Chicago at the Metallurgical Lab (particularly Szilard) offered two opposing views on the use of the bombs in Japan. As Norris notes in her own archival intervention, the former, Oppenheimer, narrative has tended to dominate, functioning as a kind of metonym for the Manhattan Project overall, but this account of the project has the effect of favoring the ex post facto moral conscience of an Oppenheimer over the more precautionary politics of Szilard.[59] Even more than Oppenheimer, Szilard was concerned with "conscientious scruples"; indeed, the iconic year 1962 witnessed not just the Cuban missile crisis and the publication of Carson's *Silent Spring*,[60] but also the founding of Szilard's anti-nuclear proliferation organization, the Council for a Livable World. And while excavating Fermi's wager (an account of which Millet includes, accompanied by the narrator's assessment that "physicists are well-known for their senses of humor")[61] or Oppenheimer's ambivalent "peril or hope" (Millet's narrator editorializes: "For a brilliant man, Oppenheimer was relying on a surprisingly impoverished logic"),[62] might offer insight into the ironies of the second nuclear age, it is Szilard's contribution to the archive that best approximates reflexivity, in part because of the genre in which he worked.

Millet's narrator describes Szilard (not entirely unjustifiably) as a "writer of exceedingly dull fictions,"[63] but he was first and foremost a writer of exceedingly dull nonfictions. Whereas Oppenheimer's writings for a nonscientific audience tended toward sweeping moralisms and philosophical generalities, Szilard's tended toward the concrete, offering specific prescriptions for how a world government might be achieved.[64] Because Szilard's solutions to the political problems of the atomic age were often so far-fetched, running "roughshod over formidable impediments—Soviet-American mistrust, cultural differences, widespread suspicion of world government, deep-rooted nationalism, powerful ideologies, and the desires of government and citizens," as Barton Bernstein notes,[65] these ended up as fictionalized narratives—indeed science fictional narratives—their programmatic elements framed by fantastical communication with dolphins, space travel, or cryogenic freezing. "The Voice of the Dolphins," which would eventually be the title story for his collection published in 1961, was originally written as an article, "Has the Time Come to Abrogate Nuclear War," a piece rejected at least twice—by *Look* magazine and then by *Foreign Affairs*—before being recast in fictional form, its recommendations now in the "voice" of the brilliant, disinterested, and rational dolphins (with whom scientists learn to speak—or the story suggests, ventriloquize).[66]

Szilard's choice to write fiction (dull or otherwise) is, then, often narrated in terms of his failure to publish nonfiction, the presumption being that a detailed program for world government would be more palatable in fictional form. But the fantastical elements of the fiction surely suggest something excessive to the mere desire to get the more mundane, if no less speculative, political plans in print. If Millet imagines Szilard, transported through time to confront himself in the archives, she is, in effect, performing a thought experiment that already appears in Szilard's own work. Whatever their differences in fictional style, both Millet and Szilard offer examples of that stock science fiction strategy of defamiliarization, to imagine a perspective—a future human being; an alien; or even one of us, transported through time via cryogenetic freeze-and-thaw—that is radically exterior to our own, one able to provide that much-needed objective position from which our own irrational behavior can be seen for what it is. In most of the stories collected in *The Voice of the Dolphins*, the narrator is a future scholar in the position of piecing together a narrative of the past based on the leavings of a partial archive. Whether it is destroyed by fire or nuclear war or time, the archive is, in each case, sketchy, in need of some fictional suturing. And in imagining the future, the archivists who will be approaching the historical "Szilard," Szilard playfully imagines his own ideas vindicated, as in the story "The Mined Cities," in which the cryogenically frozen protagonist awakens to find that "the whole sequence of events that I have just told you had been up to this point correctly predicted by Szilard in *The Voice of the Dolphins*."[67]

Indeed, the fact that Szilard was comfortable translating his visions into fiction is not wholly surprising, given that, among many other inventions, he might be said (with others) to have invented the concept of "fabulous" nuclear war. While other scientists and politicians debated whether to use the bombs in Japan, Szilard advocated an alternate solution—to bring the Japanese over to witness the Trinity test, and to invite them to surrender before any further loss of life. It is to documentation of this proposal that the main character, named "Szilard," refers in his extrapolative fiction, "My Trial as a War Criminal," which, written in 1947, plays with what Eugene Rabinowitch noted was an unintended consequence of the scientists' movement: Taking the scientists' claim that "atomic weapons are in a different class from all other implements of war," the Soviet Union (or Soviet "propaganda," as Rabinowitch judged it) argued for the "branding as war criminals of national leaders first to order the use of the atomic bomb."[68] Here, Szilard envisions a future resolution of the Cold War, in which the Soviet Union has won not with atomic weapons but with a viral bioweapon. He is then arrested for his work on the bomb and accused of war crimes in the bombing of Hiroshima. "Szilard" recalls: "I thought at first I had a good and valid defense against this latter charge, since I had warned against the military use of the bomb in the war with Japan in a memorandum which I had presented to Mr. Byrnes at Spartanburg, South Carolina, six weeks before the first bomb had been tested in New Mexico."[69] But as it turns out, this document has gone missing: "This memorandum, which Mr. Byrnes had put into a pocket of his trousers when I left him, could not be located by counsel for the defense either in the files of the State Department or in the possession of any of the Spartanburg cleaners who might have kept it as a souvenir."[70] A sly means of reinserting the memorandum into the archive, the story reminds the reader that history could have gone otherwise, even lending some credence to the "propaganda" of the Soviet Union. Ultimately, however, the Szilard character is saved, despite these gaps in the archive, when the Soviets' own bioweapon runs amok, and the system that is prosecuting him breaks down, an ironic turn of events that allows him both to remind his readers of his objections to the use of the bomb in Japan and to highlight the side effects of any weapons system.

The Soviet bioweapon might initially seem, here, akin to Vonnegut's ice-nine, a new technology that eludes the control of its makers, but whereas Vonnegut ultimately indicts both science and politics—his narrator wondering, "What hope can there be for mankind . . . when there are such men as Felix Hoenikker to give such playthings as ice-nine to such short-sighted children as almost all men and women are?"[71]—Szilard generally reserves his biting irony for politics, with science cast in the role of rational salvation. When, in "Voice of the Dolphins," the narrator describes the efficiency and effectiveness of the scientific method, he echoes Szilard's nonfictional voice as well:

Political problems were often complex, but they were rarely anywhere as deep as the scientific problems which had been solved in the first half of the century. These scientific problems had been solved with amazing rapidity because they had been constantly exposed to discussion among scientists, and thus it appeared reasonable to expect that that the solution of political problems could be greatly speeded up also if they were subjected to the same kind of discussion.[72]

Believing in the power of rational, scientific thought, Szilard felt that its application to social problems would be as fruitful as its applications to scientific and technical ones had been. Of course, the appearance of this passage in a fictional text might ironize the position in spite of Szilard's seeming endorsement of it.

The futures that Szilard imagined in his fiction varied, from the rational deterrence of "The Mined Cities" to the apocalyptic aftermath of total thermonuclear war—observed by alien visitors—in "Report on 'Grand Central Terminal.'" But none of these futures could compare, Millet's novel implies, to the world we now occupy. The fact that the scientists return in March 2003, on the eve of the second Iraq war, clearly suggests that their presence is connected to the invasion, even if Iraq is not explicitly named in the novel, and Millet is clearly responding as well to the Bush administration's Nuclear Posture Review, leaked in 2002, which called for fewer but also smaller and more usable nuclear weapons to complement the conventional arsenal. Indeed, Millet's Szilard seems to believe that it is this thanatopolitical drive that has conjured the scientists from the past: "Not just any war, but the last war. Remember the last time they said that? How hopeful it was? The War to End All Wars? But this is the real McCoy.... We're here for the last war. Because we started it. We're needed."[73] Here, the responsibility of the scientists, their power and wisdom, again appears to be called upon, but as the fictional Szilard discovers, the second nuclear age is even less amenable to their rational solutions. As Millet's Oppenheimer explains: "He thinks all rational men will automatically agree with him when he confronts them with the facts. Leo's not postmodern."[74] The world he encounters, Millet implies, is. Though her scientists are in some ways the wisest of her characters, they are also wholly ineffectual in the world their speculative imaginations produced.

The messianic end-times to which Millet's Szilard refers are indeed appropriate, for what the scientists discover in 2003 is a fairly dystopian present characterized by a toxic blend of fanaticism and complacency. In "A Secular Apocalypse," an article published in the *Bulletin of the Atomic Scien-*

tists in 2007, Sam Keen provides a characterization of Americans that would work as well to describe the world of *Oh Pure and Radiant Heart*: "Americans, it seems, are profoundly schizophrenic, split between expecting Jesus to return and simultaneously placing their hopes in a coming triumph of liberal democracy and the market. TV evangelists and Wal-Mart, fundamentalism and modernity cohabit."[75] Whether the Rapture or a Fukuyama-style "end of history," the end seems virtually inevitable in the novel, as McDonald's and Coca-Cola dominate the streets of Hiroshima and fundamentalists hire private security details.

Arriving directly from the site of the Trinity explosion, Millet's scientists also arrive with their hopes for a cosmopolitan universalism built on the terror of the new weapon fairly intact, their faith in the power of the United Nations as a body for controlling atomic power still strong. But though the world they discover in the present is perhaps more globalized, it is much less equipped for a world government. When the scientists visit Japan, they are hosted by Larry, the son of a wealthy expatriate, the seeming product of a globalization driven by capitalism, not cosmopolitan feeling. Oppenheimer addresses a group of Larry's compatriots—most of whom are marijuana-smoking followers of the Grateful Dead—only to find that his calls for internationalism fall on uncomprehending ears:

—[We] propose that all the countries of the world, both rich and poor, abandon their nationalistic fervor, said Oppenheimer . . . In short we want to propose that the United Nations, that much-maligned body, that body that has apparently been so undervalued and underused—

—United Nations? asked the man with the pigtail, confused.

—Go Oppie! shouted someone unseen, and clapping started at the back and moved forward.

—for the sake of the glorious future . . .

—This a party or what? asked the [man in the] wifebeater.

—world government. We propose a unity, not of corporations across the globe . . . but of people, of people and their directly elected representatives . . . the struggle will be long—

—Long! I can dig it, said the camo man.

—the goal of establishing—at long last—universal world peace.

—Peace, man, peace! crowed Larry, making the two-fingered sign to a terminal spatter of applause.[76]

The juxtaposition of the rational, humanist scientist and the irrational, postmodern audience is a source of humor, certainly, but of the rather bitingly sarcastic variety, making one wonder whether, after all, satire is not, as John Snyder argues, "an admission of defeat."[77] As Cathy Caruth has recently reminded us, the vulnerability of the archive is not limited to some hypothetical fiery apocalypse; the "collective loss of the knowledge of how to read" might produce a similar result.[78]

Millet's nonscientist characters have consigned the nuclear to a past that they choose not to access and do not see as integral to the present or future. When the scientists ask to visit Hiroshima's ground zero, Larry informs them that the bombing of Hiroshima is "ancient history." "No one cares about that anymore," he tells the scientists. "They barely even teach it in school."[79] In the absence of history, political organizing becomes a pastiche of past media events, culled seemingly randomly from the archives. Even the more earnest of the scientists' followers make activism look sadly anachronistic. When Ben describes the scientists' movement to Ann, he says: "We are the world. . . . Next he's going to start shooting videos for MTV. Scientists hugging each other in and swaying in front of the mikes."[80] And when the protesters perform a "die-in" at the test site, another character says, "I didn't know they even had those protests anymore. . . . I thought they went out with Ronald Reagan. But hey man, it's cool."[81]

The anachronism here is both the activist theater and its target, the bomb, for, in the post–Cold War era, while the bomb persists as mininukes or depleted uranium munitions, it is inevitably joined by the pervasive contamination of the global environment. Marking the shift from single to multiple risk, Ben recalls: "I used to have dreams of mushroom clouds when I was thirteen." "But not since then," Ann replies. "Never, said Ben.—There are too many other ways the world could end."[82] And lest we not recall what these other ways might be, Millet's narrator notes, "in recent times, bad things were happening: planes flew into buildings and democracy was waning. War was everywhere erupting and as people multiplied obscenely and advanced on open space they were driving all the plants and animals extinct"; and "the rivers and seas and the fish that swam them were flowing with mercury, forests were being felled and deserts turned into strip mines."[83] To the potential Christian apocalypse and the nuclear apocalypse, then, the novel adds the rhetoric of environmental apocalypse, which seems as inevitable as the other ends foretold.

This end-times obsession characterizes the narrative as well. Ben's constant questions to Ann about the goals of her scientists' movement seem almost metafictional. He asks, "What's the happy ending?"[84] And, even more pointedly: "When will you know when it's over? It there going to be a sign that pops up on the road and says THE END?" As if in answer to this question, Millet ultimately delivers something like a Rapture, as the scientists are lifted up from where they stand at the March on Washington and are carried off on the wings of what some read as giant cranes and some as angels. If this is the Rapture, however, essentially everyone is "left behind." Father Raymond, one of the more liberal of the religious followers, provides an interpretation: "Those were not birds, he said. . . . Or put it this way: they were not only birds. . . . They brought us a message. . . . Didn't you know? The end has already come and gone. And here we are."[85] The apocalypse moves here, as Frank Kermode famously suggested in the first atomic age, from imminence to immanence, less a spectacular bang than a whimper through which the characters, and we, must suffer for a protracted period of time.[86] The future is not as the scientists expected; the end has changed. And as they are escorted out of the novel by the giant cranes, we are left, with Ann, to query the second nuclear age.[87] Millet's reinvigorated scientists' movement and Ann's commitment to it might be read as evidence of nostalgia, yearning for an age in which nuclear weapons were "all" we had to worry about. At one point, Ann muses on her devotion to the political cause of the new scientists' movement: "Causes had always kept her at a distance: they cried out for attention but left her numb. There were just too many of them, mostly hopeless. But now there was only one."[88] At the end of the novel, with the departure of the scientists, the other causes crowd back, hopeless as ever.

But if the novel seems, in its biting satire, to view the present in terms of "peril" rather than "hope," the type of cynicism often associated with irony as trope, it may contain a more affirmative promise as well, readable less in the tone than in the relationship of the novel's content to its style. The scientists retrace the events of the half-century they have missed, visiting first the Trinity site, then Hiroshima, then the Bikini atoll and the Nevada test site, and along the way the novel gathers together what emerge as three distinct narrative modes: the first is the fictional narrative of the novel in which characters (including those based on historical figures) operate, if rather absurdly; the second is a philosophical mode, in which Ann and Oppenheimer muse to themselves on the larger questions of life and death and love and crowds; and the last is a kind of exploded historical narrative that, like an archival unconscious, erupts periodically into the narrative. Though the characters seem to have selected their own reading materials from the atomic archive—and thus some of the historical detail in the text could be explained by the device of their reading—Millet refuses to integrate these modes. Rather, each seems separate from the other, with historical facts appearing as nonsequiturs in the midst of the hijinks of the scientists and their followers. So, for example, a description of Oppenheimer being forcibly dunked in a stream by his religious followers and a description of Szilard triumphant in his attempts to get Fermi's physical body exhumed are interrupted by a single sentence: "By the turn of the millennium, nuclear weapons production facilities occupied over three thousand square miles of U.S. territory."[89] The novel itself thus reads as archive, as bits of fiction, history, and philosophy are gathered but not rendered as a coherent narrative.

Given the population that the novel describes, we might read in it some of the nihilism of Kurt Vonnegut's imagined holy man, Bokonon, who, in *Cat's Cradle*, admonishes: "Write it all down," for "Without accurate records of the past, how can men and women be expected to avoid making serious mistakes in the future?" Vonnegut's narrator, John, clarifies the tone: "He is really telling us, of course, how futile it is to write or read histories."[90] As if responding cynically to Beck's "enlightenment function" in risk society, Millet's Ann wonders whether "there [is] a difference between waiting for enlightenment and waiting to be entertained."[91] But while Vonnegut's postmodern status may be up for debate, I would submit that Millet is, like her Szilard, not particularly postmodern, her use of something like historiographic metafiction notwithstanding, for she too, in her recounting of history—from the number of above-ground blasts at the Nevada test site, to a description of Project Plowshare and the Aleutian islands testing, to the statistics of weapons stockpiled by China and the Soviet Union, to a list of the signers of the Comprehensive Test Ban Treaty of 1996—seems to hope that "all rational readers will automatically agree with her when she confronts them with the facts," even if such rational readers are few and far between in the novel itself. If there is a revelation in the novel, it comes neither in the fictional apocalypse, nor in the wisdom of the returning scientists, nor on the sweeping wings of giant cranes, nor even in the kind of consumerist politics that Beck has elsewhere suggested as a possible solution (indeed, food activists in particular are pilloried in the novel); rather, the "enlightenment" comes in the productive gap between the accumulation of historical details of the environmental and social impacts of the first nuclear age and the seeming inability of the inhabitants of the second nuclear age to comprehend it, a gap that presumes a reader, a "risk critic," capable of an as yet unimaginable integration.

Given its comedic focus on the three atomic scientists, *Oh Pure and Radiant Heart* reads, in part, as satire of the narrative of the atomic age that Ken Cooper describes as fetishizing the "ground zero" of Trinity, the alpha and omega of the nuclear age. Indeed, Millet's novel does much to reinforce (if potentially ironically) the "whiteness of the bomb," for, despite the scientists' desire for cosmopolitanism and despite their actual travel to Japan, the people they encounter are almost exclusively American (and generally racially unmarked, which often implies whiteness). And though there are references to the historical unevenness of the bomb—the testing in the Pacific, for example—the scientists' visits to Bikini are more like diving adventures than attempts to measure the bomb's effects. Though the scientists begin in poverty (their access to bank accounts complicated by the fact of their deaths), they quickly pull themselves into financial stability, first by sapping Anne's lifesavings and, later, by relying on Larry the trust-fund hippie's substantial resources, leaving them still in the Olympian position of relative privilege. Indeed, missing altogether the ways in which inequalities might affect vulnerability to toxic threat, the scientists feel that they themselves are persecuted, in this case by what they perceive to be the misplaced risk management that sees smoking as a danger to public health. Oppenheimer's struggle to survive a transpacific plane ride on nothing but nicotine gum provides some of the comic relief in the novel.

Read in the context of Szilard's work, however, the scientists' interpretation of their own persecuted status as smokers offers something more than comedy, as it presents an inversion of a scenario imagined in one of the stories from *The Voice of the Dolphins*, "Report on 'Grand Central Terminal,'" in which two alien characters discover a postapocalyptic New York City. Observing signs for "smoking" and "nonsmoking," as well as images of people of different hues, these extraterrestrial observers presume that these signs refer to "smokey" people rather than to any particular act; they are thus correct in their identification of racial segregation, but comically incorrect in their linking of those practices to smoking. In the perhaps more subtle context of the present, Millet's imagined scientists are arguably as confused as Szilard's extraterrestrials, their alien focus on smoking largely obscuring the kind of environmental justice issues that someone like Toni Cade Bambara labored to present.

Oh Pure and Radiant Heart seems to mark the messianic atomic scientist as an anachronism, but arguably this model of the scientist as general intellectual persists, if somewhat tempered, in the second nuclear age. When the *Bulletin of the Atomic Scientists* declared the start of the "second nuclear age" in 2007, the movement of the Doomsday Clock attracted the interest of the news media, and this provided a formal occasion for contemporary scientist-celebrities to enter the public sphere to speak as "experts" on the perils to come, perils that, like the nuclear bomb, science itself has had a large hand in developing, whether in genetic modification or nanotechnology, carbon production or toxic contamination. These pronouncements on the occasion of the second nuclear age recalled, fairly remarkably, those that marked the inauguration of the first. Combining the language of peril and hope that characterized the scientists' movement in the 1940s, Stephen Hawking asserted:

> As scientists, we understand the dangers of nuclear weapons and their devastating effects, and we are learning how human activities and technologies are affecting climate systems in ways that may forever change life on Earth. As citizens of the world, we have a duty to alert the public to the unnecessary risks that we live with every day, and to the perils we foresee if governments and societies do not take action now to render nuclear weapons obsolete and to prevent further climate change.[92]

The call to internationalism in the face of immanent peril, the sense of the urgency of the timing—"action now"—and the fairly utopian language—"render nuclear weapons obsolete"—all recall those statements issued by the early atomic scientists, as though Hawking were a kind of resurrection of Oppenheimer or Szilard, come to warn us of the end of days. In this context, Millet's farcical scientist movement suggests the importance of interleafing this scientific messianism with a historical consciousness. When, in *Oh Pure and Radiant Heart*, the media interview one of the participants at the scientists' peace rally, "Ron Subac, a tenth-grader from Reno," a "teenager wearing a T-shirt with a line drawing of Oppenheimer in his porkpie hat," the interviewer asks him whether he believes that these scientists are in fact resurrected in the flesh from the end of World War II. Ron replies, "Totally! Time travel! I mean I read Stephen Hawking. Do you?"[93] Ron's query to the interviewer—"Do you?"—might be taken as an invitation to Millet's reader to read differently, reading Hawking back into the context of the first nuclear age, not for the inane purpose of entertainment, but for the more serious purpose of archival reflections.

The precedent of the scientists' movement may inject some doubt into the prospects for Beck's own risk cosmopolitanism, which is also driven by what he believes is the primary emotion that characterizes risk society: fear. As Rabinowitch noted, fear can easily be mobilized in the service of war rather than peace (perhaps more easily). And, in an age of risk, scientific warnings of impending doom can become so much background noise for a public inundated by crisis culture. But, of course, Beck's subpolitics departs from the scientists' movement in one key respect: for Beck, the politics of risk comes not as a result of the advice of scientific experts, but as a result of the failures of scientific expertise, a context perhaps hinted at in Hawking's scientists who are "*learning* how human activities and technologies are affecting climate systems." Beck writes: "The sciences are *entirely incapable* of reacting adequately to civilizational risks, since they are prominently involved in the origin and growth of those very risks. Instead—sometimes with the clear conscience of 'pure scientific method,' sometimes with increasing pangs of guilt—the sciences become the *legitimating patrons* of a global industrial pollution and contamination of air, water, foodstuffs, etc."[94] It is only in the "reflexive" aspects of second modernity, the confrontation of science with itself, its own side effects now seen as effects, that the "enlightenment" function might produce the ironically dialectical cosmopolitan future.

ORIGINS, IN HINDSIGHT

The atomic scientists at midcentury certainly considered the risks and benefits of nuclear energy, for weapons and for power. The pragmatic concern that prompted Szilard and Einstein to write to FDR was the immediate risk that the Nazis might obtain this technology, a risk that, in hindsight, appears to have been overblown. The scientists' movement after the war projected a hopeful future for the atomic archive, a narrative of risk in which the bomb would mark the commencement of an age of cosmopolitan peace, a projection that, today, reads as part tragedy, part farce. Had that future come to pass, Oppenheimer's commandment might have been an apt arché, a recognition of responsibility in the face of the potential power of annihilation, but in the second nuclear age, the age of side effects and accidents, of leaks and spills, of climate change, with its unanticipated surprises, new ends lead to different beginnings.

Millet's scientists' transport from the Trinity test to the present seems to indicate, as the scientists themselves speculate, that there is "something in the event, something anomalous and unprecedented" that made their time travel possible.[95] And certainly, as the first demonstration of the destructive power of the bomb, Trinity does seem a fitting commencement for a nuclear age overshadowed by the "fabulous" promise of a total thermonuclear war. But, as critics like Ken Cooper (focused on historical precedent) and Elizabeth DeLoughrey (focused on toxic legacies) have suggested, this model of the first nuclear age is itself partial, for the blast of the weapons risks blinding us to the less spectacular consequences of the nuclear age: "The shock of an eventist model of history . . . should not distract our attention from the impact of a longue durée of radiation ecologies, particularly when we consider that nuclear weapon byproducts, such as carbon-14 and plutonium-239 have 5,700 and 24,000 year half-lives."[96] In the second nuclear age, in fact, DeLillo's meeting of weapons and waste in *Underworld* marks not the end of an era but the beginning of the next, as depleted uranium munitions, used in the Gulf wars, offer a new use for nuclear waste, which as Peter van Wyck has suggested, one could read as "an attempt at an industrial-military plus-sum solution to war and waste" insofar as it is a kind of "recycling" of waste associated with nuclear power production that also clearly "wastes" the areas it contaminates.[97] If "mini-nukes" and depleted uranium, climate change and biotechnology, ozone holes and fire retardants are the new hazards ushering in a protracted apocalypse, then perhaps the spectacular explosion at Trinity is not the model arché for our new telos. And even Fermi's wager, which Jasanoff identifies, in hindsight, as the event that inaugurates global risk, may not be properly enlightening, for, given the unpredictability of risk, given its incalculability, the true wager in risk society is always the bet you don't quite know you are making.

Just a few years after Trinity, when the atomic secret was out, David Bradley, a doctor working with the radiological unit in the Pacific testing grounds, would record a different set of wagers; this time, given that the bombs were bigger and were exploded in water, new unknowns arose and new apocalyptic scenarios were imagined:

> This was Able Day. . . . Would the prophecies they had read and heard be fulfilled—would pieces of our venerable Navy be spread all over the Pacific from the Philippines to Panama? Would a tidal wave sweep the islands clean and surge on to inundate Los Angeles? Would, indeed, the very water itself become involved in a chain re-

action until the whole Pacific Ocean disappeared in a colossal eruption? Who was to say? Or who, at least, was to say No?[98]

Given the unprecedented nature of the event, Bradley's closing questions were, of course, rhetorical. No one could guarantee the outcome, and the explosions proceeded in the face of this uncertainty. After the tests, Bradley reports, there was some disappointment when none of the apocalyptic scenarios came to pass. But what the observers, Bradley among them, discovered later was that the risk was not only in the fiery explosion, the threat of total, spectacular destruction, but something more subtle, persistent, and insidious, as the fallout spread on the wind to poison those on nearby atolls and fishing vessels.[99] The navy returned to the ships in an attempt to scour them of the dangerous and invisible radioactive particles, only to discover that there was simply no means of removing them. Bradley describes the scene: "decks you can't stay on for more than a few minutes but which seem like other decks; air you can't breathe without gas masks but which smells like all other air; water you can't swim in, and good tuna and jacks you can't eat. It's a fouled-up world."[100] Incalculable risk is indeed incalculable—not only because, as Jasanoff says, risk is "probability of the harm times the magnitude of the harm"[101]—but because "harm" itself may not be knowable in advance, and such calculations are therefore impossible.

In this context, in the spirit of Jasanoff's intervention, I would like to turn the clock back still further, before the explosion at Trinity, before Fermi's wager, a moment suggested in Eugene Rabinowitch's postmortem on the scientists' movement, "Five Years After." There, Rabinowitch opens with two events, both of which, he suggests, might have ignited the spirit of the scientists' movement:

> Standing around the first fire lit under the West Stands of the Athletic Field of the University of Chicago in December of 1942, and two-and-a-half years later, in July of 1945, watching the flash of the first atomic bomb explosion here at Alamogordo, the scientists had a vision of terrible clarity: They saw the cities of the world, including their own, falling into dust and going up in flames.[102]

The second event is, obviously, the Trinity explosion—and that the scientists might have foreseen the apocalyptic military potential of the weapon at that point certainly seems plausible. The first, however, is the atomic pile, Fermi's primary contribution to the project, an event that proved that a chain reaction could be sustained; and here the relationship of event to prophetic apocalyptic future is less clear. In an article titled "Fermi's Own Story," published in the *Chicago-Sun Times* in 1952, Fermi recalled: "The sequence of discoveries leading to the atomic chain reaction was part of the search of science for a fuller explanation of nature and the world around us. No one had any idea or intent in the beginning of contributing to a major industrial or military development."[103] Certainly, there is something disingenuous about Fermi's claim not to have such foresight. Though his own motivations may have been scientific, he certainly knew his pile was destined for just those developments. As he notes elsewhere, "Through the collaboration of all of the men of the Metallurgical project and of the Du Pont Company, only about two years after the experimental operation of the first pile large plants based essentially on the same principle were put in operation by the Du Pont Company at Hanford, producing huge amounts of energy and relatively large amounts of the new element, plutonium."[104] The participation of DuPont clearly might have tipped Fermi off to the potential for "industrial development."

But important as this future of the atomic pile is to the age of risk, I would like to stay for a moment in the event of the pile itself, an event that had, arguably, packed into it a range of possible futures, but that also had a material presence the future of which was as uncertain. Working with Fermi on the pile was a young physicist named Leo Seren, a man who would later, as Szilard's imagined alter ego never quite did, call himself a "war criminal" for his work on the bomb.[105] Interviewed fifty years after the bombing of Hiroshima, Seren described the source of his sense of complicity:

> "When the piles went critical, neutrons were escaping into the neighborhood," Seren said. "One of my duties, after Dec. 2, 1942, was to put neutron detectors up to a block away in apartment buildings in Hyde Park."
>
> The levels escaping into the neighborhood from unshielded piles, he says, were not a major danger to health. But one problem with unwanted radiation, he said, is that "it is cumulative. It builds up with detrimental effects."[106]

My point here is not to make some revelation about the risks faced by residents of Hyde Park. Those residents of the neighborhoods of Hiroshima and Nagasaki, of the Pacific and Nevada testing grounds, were—and continue to be, considering the persistence of radiation—clearly at much greater risk. But the image of the unshielded pile on the University of Chicago campus, secretly leaking its radiation into the surrounding areas, pro-

60 RISK CRITICISM

vides nonetheless an apt figure for the wagers of the risk society: the uncertainties, the additive and cumulative qualities of what might seem to be minor or isolated hazards.

THE CRITICAL WAGER: ACCUMULATING RISK (ARCHIVES)

As Ken Cooper's critique of the "Olympian" story of the bomb suggests, archiving is itself an act of interpretation. Beginning with, say, Oppenheimer and Trinity produces a particular story of the atomic age. Oppenheimer's own wager on the future was, perhaps, on irony, as the bomb was to bring peace, but irony is a slippery trope; it seldom proceeds according to plan. Cooper's contribution to thinking the bomb is to supplement the Olympian story, expanding the archive beyond its usual bounds, and in the process transforming it. Supplementary work also characterizes the texts I have addressed here, this time the expansion coming in the form of the multiplication of hazards, suggesting the cumulative nature of global risk. In this context, the atomic pile surely provides an example and an ironic origin, but it may also offer a figure for the practice of risk criticism itself. Indeed, in returning to the pile, I am once again following and supplementing the work of Bruno Latour, who, finding a paucity of appropriate metaphors for contemporary critical practice (rejecting the "neutron bombs of deconstruction" and the "missiles of discourse analysis"), returns nonetheless to the atomic archive in order to repurpose Fermi's pile for a new context.[107] In place of the anachronistic and destructive vehicles (and the critical tenors he deploys them to illuminate), we need, he asserts, a "whole new set of positive metaphors," now in the service of a practice capable of grappling with the hybrid actants, the "things" in the sense of "gatherings" that ought, as I too believe, to animate our work.

As a start, he turns to "a most unlikely source," Alan Turing's 1950 essay on thinking machines. Turing is there discussing computers, and the question in the original article is whether one might get something more out of a computer than one put in it. Reaching for a metaphor through which to understand this question, Turing first turns to a piano string, which indeed, "struck by a hammer," "will respond to a certain extent and then drop into quiescence."[108] Without further input, there is no further output. But for Turing's thinking machine another analogy is in order, and for this he draws from the other technological marvel of his time, the atomic pile:

> Another simile would be an atomic pile of less than critical size: an injected idea is to correspond to a neutron entering the pile from with-

2. Chicago Pile I (CP-I): World's First Reactor. (Courtesy of Argonne National Laboratory.)

out. Each such neutron will cause a certain disturbance which eventually dies away. If, however, the size of the pile is sufficiently increased, the disturbance caused by such an incoming neutron will very likely go on and on increasing until the whole pile is destroyed.[109]

Deploying the pile as metaphor, Turing goes on to point out that the same might be said for human minds—some are "subcritical," seeming to respond to input with limited output, while others seem "supercritical," input "giving rise to a whole 'theory' consisting of secondary, tertiary and more remote ideas," and he asks whether this latter might be possible for machines.[110] Ferrying this metaphor over into his discussion of critique, Latour asks: "What would critique do if it could be associated with more, not with less, with multiplication, not subtraction. Critical theory died away long ago; can we become critical again, in the sense here offered by Turing? That is, generating more ideas than we have received, inheriting from a prestigious critical tradition but not letting it die away, or 'dropping into quiescence' like a piano no longer struck."[111]

Neither Turing nor Latour is very interested in the materialities of the source for this vehicle—the atomic pile itself. They are focused instead on the tenor they hope to communicate, the idea of limited input and extensive output—of the multiplication of information or thought. But considering the degree to which Latour's argument centers on the multiplication, the gathering, the "thing-ness" of things, the cleansing of this metaphor of its radioactive content is itself fairly ironic. Fermi's pile indeed operates by producing "more," by "multiplication not subtraction," but what is multiplied is not confined to the pile itself, as the radiation produced accumulates in the bodies around it, potentially inducing a veritable proliferation of cells. And one could certainly emphasize the end point of the critical pile, which, after the initial excitement of particles, is, after all, destroyed, leaving only dangerous and useless by-products, hardly the model for sustainability, in environmental or critical practice. Scientists may indeed, as C. P. Snow once said, have the "future in their bones," but in the case of the atomic scientists, this is as literal as it is metaphorical.[112] Fermi died, like so many of the atomic scientists, of cancer.

Have we now reached the origin of the world at risk? The place where it all began? Of course not. The pile is a figure, but it is a figure that must, for the becoming-real of risk, be historicized, materialized. Latour mobilizes it against "subtraction," the ways in which critique takes us further away from the real, but in the very process of metaphorical transfer he too subtracts materiality. By all means, I agree with Latour: Let us beat our critical arsenals into more fruitful plowshares. Beck's model of risk society, too, demands something like addition or multiplication over subtraction; this is the logic of the archive of risk. Risk society is characterized by the accumulation of otherwise seemingly isolated events, past, present, and to come—melamine poisoning and thalidomide, asbestos, mercury, and BPA. It is only by seeing in these risks and catastrophes a common underlying structure that renders them all analogous that a "world risk society" can be envisioned—an archival process that, for Beck, is crucial to the development of the dialectical possibilities for ironic reversal of a world at risk. But such a critical approach must also be attentive to its own dangers. The risk critic is here no messiah—we cannot, as Oppenheimer did, bring the "good news" of peril and hope—or at least not without irony. Viewed as a figure for risk that a "risk criticism" might take as its object and take as a model for what might be placed in the archive, then, Fermi's pile is indeed potent—and multiplication an apt practice—but only when the vehicle retains the traces of its material origins. Recalling the pile, Fermi noted: "The event was not spectacular, no fuses burned, no lights flashed. But to us it meant that release of atomic energy on a large scale would only be a matter of time."[113] We are still grappling with the consequences of that release over time, what we might, following Rob Nixon, call the "slow violence" of the second nuclear age.[114] As Millet reminds us, "By the beginning of the twenty-first century the men who had been central to the design and construction of the atom bomb a half century earlier were dead. The bombs they had conceived remained, of course; the bombs in their various silos, trucks and trains, their submarines and aircraft, had been dispersed over the globe like seeds, and lay quietly waiting to bloom."[115]

Though the anticipations of the first nuclear age might have conditioned us to think of this "blooming" as the apocalypse, that unimaginable future without a future, the second nuclear age offers something much less total. We do not know what the future will bring—nor how that future will change our present or past. What will the appropriate boundary have been for the archive of the nuclear age? A fitting speculative model might come from Bambara's *The Salt Eaters*, which closes with a rumbling that might be thunder or an explosion at the "Transchemical" plant or a total thermonuclear war or something else altogether ("whatever cataclysmic event it might turn out to be").[116] The novel's final pages offer a speculative fiction in the form of a flash-forward to what might be to come, "many years hence, when 'rad' and 'rem' would riddle everyday speech and the suffix '-curies' would radically alter all assumptions on which 'security' had once been built."[117] Here, the narrative voice offers a reflection on the impossibilities of boundaries, whether of time or space, geography, body, or archive: "No one would say 'across the border,' for that entailed tiring explanations, obliged the speaker to be precise about what border was meant—where Legba stood at the gate? Where Isis lifted the veil? The probable realms of impossibility beyond the limits of scientific certainty? The uncharted territory beyond the danger zone of 'safe' dosage? The brain-blood or placenta barrier that couldn't screen plutonium out?"[118] The use of the conditional tense—"would"—offers a precautionary vision of the future, not as a certain total fiery (or frozen) apocalypse, but as a world of risk, the breakdown of boundaries producing a proliferation of incalculable unknowns, despite the lay familiarity with rad and rem. The world here is most certainly "one," and the boundaries and borders with which we are familiar may no longer be the ones most vital to navigate. But as the novel that leads up to this speculative future moment reminds us, the legacies of those older boundaries—of nation, race, class, gender—continue to divide in a world at risk.

TWO | We All Live in Bhopal?
Staging Global Risk

> *Love Canal, Three Mile Island, Bhopal, Chernobyl, the Exxon Valdez: this modern mantra lists both actual incidents and their subsequent history in the postindustrial imagination that have ensured that the environmental apocalypticism triggered by Hiroshima and Nagasaki would outlast the cold war.*
> —Lawrence Buell, "Toxic Discourse"[1]

> *We're all downwinders now, some sooner than others.*
> —Rob Nixon, *Slow Violence and the Environmentalism of the Poor*[2]

WHO LIVES IN BHOPAL? A RIDDLE FOR RISK SOCIETY

In *No Place to Hide*, his firsthand layman's account of the atomic testing near the Bikini atoll in 1946, David Bradley makes a rhetorical move that might be said to be emblematic of world risk society.[3] "The Bikinese," he explains, "160 odd people, are not the first, nor will they be the last, to be left homeless and impoverished by the inexorable Bomb. They have no choice in the matter, and very little understanding of it. But in this perhaps they are not so different from us all."[4] In short, as he says in the introduction, "Bikini is our world."[5] Meant to capture the global reach of atomic risk, the move here is necessarily a displacement, something like synecdoche, for clearly Bikini is a part of the world now standing as its whole, but perhaps closer to a double movement, first metonymy—"Bikini" become the sign for the horror of destruction and toxic fallout wrought by Able and Baker—and then metaphor, this radioactive wasteland become the world at large. Clearly there is something hyperbolic here and, in the process, a diminishment of the particular, catastrophic experience of the Bikinese. But this was hardly unique in writing about the bomb. The purpose of nuclear testing was, after all, to show in miniature what might happen at a larger scale. More striking, perhaps, is the tense of Bradley's pronouncement. This is not the move of so many commentators to make nuclear events of the past into speculative models of the future. It is not, as in Jonathan Schell's mobilization of Hiroshima, to say "it is a picture of what our whole world is always poised to become."[6] Rather, Bikini *is* our world, a world in which radiation, invisible to the naked eye and absolutely deadly for life itself, is potentially everywhere.

Some forty years later, another commentator would make a similar gesture with reference to another seemingly local event. "We all live in Bhopal," proclaimed the headline of an article by George Bradford (a pseudonym for activist David Watson) published in *The Fifth Estate* in 1985.[7] As in Bradley's Bikini, the "Bhopal" in which we all live is one brought into the global spotlight by explosive catastrophe. In this case, not the intentional test of a nuclear weapon but an accidental explosion of highly toxic gas, methyl-isocyanate (MIC), used in the manufacture of carbamate pesticides like Sevin, Larvin, and Temik.[8] Bradford does highlight the particularity of Bhopal, where the presence of the Union Carbide plant was part of the green revolution and its accompanying chemical cocktail, but, for Bradford, in an era of pervasive toxins, when chemicals of this sort are routinely "vented into the air and water, dumped into rivers, ponds and streams, fed to animals going to market, sprayed on lawns and roadways, sprayed on food crops, every day, everywhere," the sudden explosion of gas in central India was just the most obvious manifestation of a less visible but no less insidious poisoning of the planet more generally. And when he suggests that the "dramatic" event of the explosion "almost comes to serve as a diversion, a deterrence machine to take our mind off the pervasive reality which Bhopal truly represents," he makes of Bhopal a rhetorically useful metaphor that may also empty the event of its specific historical, geopolitical, and toxicological content. Closing with a gloss on the "global village," Bradford notes that the Bhopali phrase "go to the village" means to escape danger, and he asks: "What are we to do when there is no village to go to? When we all live in Bhopal and Bhopal is everywhere?"[9]

How does one experience "global risk"? After all, catastrophe in risk society does not seem to happen everywhere all at once—at least not in the way that those projections of total thermonuclear war seemed to promise. It happens here—in this incidence of mercury poisoning—and there—in that toxic spill. It is in this cancer cluster or that record heat, this reactor meltdown or that bird die-off. A sense of global risk is, as Leo Seren reminded us of radiation exposure, cumulative, whether of a single threat, say, in the increasing awareness of the impact of atmospheric testing, or in the addition of multiple, different risks, as in Buell's litany, "Love Canal,

Three Mile Island, Bhopal, Chernobyl, the Exxon Valdez"—a list that combines chemical toxins with radioactivity and petrochemical spill to produce a broader sense of the "postindustrial imagination." And this is one way to read these rhetorical moves—radiation is not only in Bikini but also here; chemical pesticides poison not just Bhopalis but also us.

As hazard moves from the futurity of the nuclear threat—at least as it was conceived by the atomic scientists—to a kind of secret presence that is all around us, unseen and unacknowledged but not less present, the "decks you can't stay on for more than a few minutes but which seem like other decks; air you can't breathe without gas masks but which smells like all other air,"[10] the substitution of a local place of known devastation for the global condition offers one possibility. The revelatory present tense—"We all live in Bhopal"—blurs the distinction between risk and catastrophe, staging the world at large as characterized by risks in the process of "becoming-real," disasters that are happening, but not yet part of the public consciousness. And this is arguably what sets the anticipatory nature of Cold War nuclear risk apart from the staging of other sorts of megarisks. Though Ulrich Beck emphasizes that "risks are always *future* events that *may* occur, that *threaten* us," and thus the staging of risk is supposed, like the nuclear holocaust, to prevent catastrophes from coming, his example is "the debate on climate change which is supposed to prevent climate change."[11] As many commentators have argued, however, climate change is at once a future threat and an already occurring catastrophe; regardless of the effects a more convincing or terrifying narrative of future catastrophe might have, the damage already will have been done, even if we don't see its full extent at present.

In parsing the rhetorical effects of these slogans, we might draw upon the distinction between metaphor and metonymy provided by Paul de Man, who aligned these tropes with "necessity and chance,"[12] respectively. De Man's model is, at first glance, a way to think about the metonymy of, say, Bhopal, which became "Bhopal" by virtue of the accident, a seemingly random or chance occurrence. Mobilized then as metaphor, Bhopal becomes necessity, revealing the secret truth of the planet as a whole, itself in the throes of an accident so universal that it defies perception. The tropological wager is that accident might reveal necessity. But to see the metonymy in Bikini and Bhopal as pure "chance" would be to miss the "necessity" involved in the original metonymy, for the seeming contingency of these geographical sites disguises what amounts to an outsourcing of U.S. risk. In the case of Bikini, this is fairly explicit, for the tests themselves were hardly accidental, even if they produced unexpected consequences; in the case of Bhopal, though the explosion was accidental, it too had elements of necessity—or at least predictability—revealing the ways in which multinational corporations like Union Carbide played the unevenness of the global market by capitalizing on the gaps in risk regulation. "Bikini is our world" and "We all live in Bhopal" are thus in part attempts to repatriate the risks that might otherwise seem far distant—or at least to suggest that outsourcing catastrophe does not work. The problem comes, of course, in the inevitable reduction that these metonymies effect, as the "necessity" of the latter metaphorical move serves to heighten the sense of "chance" in the former, obscuring what made Bikini or Bhopal particularly vulnerable to this sort of catastrophe. And if the clearly horrific consequences of atomic fallout and MIC gas are here deployed as vehicles to heighten our sense of global environmental devastation, the very scale of the tenor has the effect of diluting the vehicle, if "our" experience is now basically revelatory of "theirs" as well.

The effectiveness of "we all live in Bhopal" is precisely in its counterintuitiveness, its suggestion that those risks, borne normally by other people elsewhere, might be affecting "us," "here," what Ulrich Beck calls the "boomerang effect" of global risk. "Bikini is our world" and "We all live in Bhopal" thus might come to stand for the ambivalence in thinking the generality and specificity of global risk, an ambivalence that notoriously characterizes Beck's work: On the one hand, "We are all trapped in a shared global space of threats—without exit," and, on the other hand, we inhabit a "world in which both wealth and risks are radically unequally distributed."[13] Bhopal may be closely associated—and forever impacted—by the Union Carbide disaster, but to collapse the two is to reduce the complexities of place and time that make "Bhopal" less portable as metaphor. Even if pesticides have become pervasive, even if strontium 90 can be found in baby teeth worldwide, risks and the catastrophic effects of failures in risk management are clearly unevenly distributed.[14]

Navigating questions of particularity and universality has, of course, been a central concern of literary and cultural studies more broadly in the last twenty or so years, as endorsements and critiques of globalism, cosmopolitanism, eco-cosmopolitanism, and the like have dominated critical conversations. In the field of ecocriticism, a privileging of the local—often linked, too quickly, to an "earth-from-space" "sense of planet" (to borrow a phrase from Ursula Heise)[15]—has, under pressure from postcolonial theorists, gradually given way to a more nuanced engagement with diverse spaces, often in the process contributing to a transformation of the "environments" under consideration, as the pastoral and the pristine have been complicated by the urban and the toxic. Engagement with environmental justice has made ecocritics more wary of easy generalizations about the "global environment," such that "all in the same boat" environmentalisms have been moderated by attention to global inequities.

It is clearly important, then, to remind ourselves that we don't "all live in Bhopal"—but it is also insufficient. After all, pesticides *are* pervasive, "vented into the air and water, dumped into rivers, ponds and streams," and to articulate these as of a piece with catastrophes around the world can be a powerful invocation of solidarity. In this way, "We all live in Bhopal" resonates with the work of Bhopali activists who have repeatedly emphasized that the event is not singular or particular. In a pamphlet covering the Bhopal accident, entitled *No Place to Run* (1985)—an uncanny echo of Bradley's *No Place to Hide*—the authors emphasize that "what has given Bhopal such an impact in the rest of the world is the suspicion that these factors may not be unique to this one disaster. . . . We should all be concerned: Bhopal is a lesson for us all."[16] Indeed, the student group International Campaign for Justice in Bhopal echoed Bradford's slogan in a statement posted on its website in 2013, proclaiming that the "North American Advisory Board (AB) stands by the campaign slogan, 'We all live in Bhopal.'" "By this," the statement clarifies, "we mean that, in a world in which corporations operate with impunity, the chemical tragedy that happened in Bhopal could happen, and in fact is happening, even in our own communities and work-places. The AB strives to educate the world of this fact by carrying the story of Bhopal and its survivors into the global consciousness, in an effort to inspire action and change."[17] If the metonymic reduction of Bhopal is not to the danger of MIC—or even pesticides—particularly, but of any sort of toxin about which laypeople (and to some degree experts as well), like Bradley's Bikinese, know nothing; if "Bikini" comes to stand for willingness to take incalculable risks or "Bhopal" the callousness of corporate internalizing of profits and externalizing of hazard, these places can become nodal sites in a chain of analogical experiences in world risk society.

The question "Who lives in Bhopal?" is thus simple enough, but "Who lives in 'Bhopal'?" is more complicated, provoking questions about the politics of metonymy, metaphor, and analogy in a context of global environmental staging, whether in the media, Beck's privileged locus for mobilizing a risk public, or in our own critical practices. For Beck, the media, in the best of cases, function as a way for activists to publicize environmental ills or corporate or government corruption in order to mobilize the viewing public. The fact that both "Bikini" and "Bhopal" could stand as shorthand for the catastrophes that befell them speaks to the fact that both were "staged" for U.S. consumption and thus were available for metaphoric extension. In an optimistic response to his own critical question—"Who discovers (or invents), and how, symbols that disclose the structural character of the problems while at the same time fostering the ability to act?"—Beck cites the example of Greenpeace, which, in its staging of theatrical protest, organized a successful transnational consumer boycott of Shell in the mid-1990s.[18] Restaged in this context in Beck's theorizing, the Greenpeace boycott indeed takes on a heroic role, the emblematic example of the subpolitical possibilities of risk society. But if the new public sphere is the media and the "votes" are dollars, access to political power is clearly unevenly distributed, and those most at risk are also those least likely to capture media interest or participate in boycott democracy. Further, as Rob Nixon has recently persuasively argued, the media, with their preference for the flash of spectacle and their remarkably short attention span, are a problematic vehicle for the communication of the "slow violence" of environmental catastrophes, whether that is in the long-term effects at Bikini or Bhopal or the more nebulous and diffuse spills and leaks of radiation or toxic chemicals that characterize the world more generally. Given the "abstractness and omnipresence of destruction which keep world risk society going," Beck notes, "tangible, simplifying symbols, in which cultural nerve fibres are touched and alarmed, here take on central political importance."[19] But, as one spectacle quickly replaces another, the kind of "tangible, simplifying symbols" that "touch and alarm" "cultural nerve fibres" in the media are seldom subject to the careful thought that might lead to meaningful political action.

Such careful thought is, of course, the job of the risk critic, the scholar who, unlike the activist faced with the more pragmatic question of how to lure the cameras, has the luxury of reflection, even as we are perhaps as invested, ultimately, in representing the "symbols that disclose the structural character of the problems while at the same time fostering the ability to act." In the interest of grappling, then, with the questions of particularity and universality, of media access and political action in the world risk society, I will turn here to two literary texts that both themselves stage these issues and that have, in turn, been staged, by critics, as touchstones representative of what Lawrence Buell calls "toxic discourse," Don DeLillo's postmodern classic *White Noise* (1985) and Indra Sinha's more recent *Animal's People* (2007), two novels that depict both catastrophes and their staging in the media, thereby raising questions about the efficacy of a global boycott democracy. Set in two very different locales—the United States and India—these novels present characters grappling with toxic exposure, and thus offer an opportunity to explore the ways in which risk and catastrophe might be experienced differently. But there is something even more specific that draws these texts together—their relationship to Bhopal, an event that *White Noise*, published just a month after the disaster, was said to anticipate, and that *Animal's People*, published some twenty years later, represents explicitly. Reading these texts' staging of risk and the ways in which they have, in turn, been staged in the criticism, offers a kind of test case for thinking through the universality and specificity of risk—and the rhetorical mechanisms by which a larger archive of analogous risks might be assembled.

BHOPAL'S SISTER CITY?

In his recent addition to the already fully stocked larder of DeLillo criticism, Matthew Packer asks: "Has Don DeLillo's supermarket satire, *White Noise*, passed its own use-by date?"[20] His anxiety here is less that the novel is somehow archaic or passé, a relic of an earlier age, than that the novel has been so thoroughly "used" by other critics. A satire of the simulacral logic of risk management on the part of experts and the complacent consumerist escapism and apathy of laypeople, *White Noise* has become canonical for a diverse range of contemporary literary and cultural criticism—whether one is interested in media or postmodernism or whiteness or globalization.[21] Such abundance makes it difficult to read the novel anew; indeed, invoking one of the most-read scenes in the novel, protagonist Jack Gladney's encounter with the "Most Photographed Barn in America," Packer suggests that Jack's colleague's assessment—"No one sees the barn"—might be applied as well at this point to the novel itself, which, so framed by Baudrillardian precession of the simulacra, has become unreadable. In the ecocritical context, the emblematic scene might be one that appears somewhat later, when Jack and his wife, Babette, contemplate the notion of a microorganism that has been genetically engineered to consume the toxins in the cloud of pesticide by-product that has blanketed the city. Babette explains the source of her "vague foreboding" by querying, "Have they thought it through completely?"[22] The answer, implied in the perpetual failures of risk management in the novel, is a definitive no; no one seems to have done the extrapolative imagining necessary to forestall the seemingly inevitable ironies of situation that are bound to follow, a fate foreshadowed in the ominous acknowledgment: "No one knew what would happen to the toxic waste once it was eaten or to the microorganisms once they were finished eating."[23] Here too one is tempted to use the novel as a vehicle for summing up the criticism: *White Noise* clearly depicts a failure to think it through completely, but the novel itself, including its representation of media, risk, and the environment, has been fairly thoroughly thought. Indeed, with an "airborne toxic event" at its ironic center, *White Noise* has taken on a kind of iconic status in ecocriticism in general and in the budding field of "risk criticism" more specifically, as representative of an ecology (whether "postmodern" or "toxic") no longer tied to conventional notions of "nature."[24]

I would like to suggest, though, that this iconic text—and its staging in reviews and criticism—may have still more thinking to inspire, particularly as a result of the ways in which critics have contextualized it in terms of what amount to analogical real-world historical events that are to correspond to the "airborne toxic event" at the novel's center. There is, of course, no specific historical referent for DeLillo's imagined catastrophe, and its very openness, its description in terms of simulacra or the sublime, is what, for Lawrence Buell, makes it less appropriate as "toxic discourse," less politically engaged than, say, Carson's *Silent Spring*: "It is hard not to conclude that a very different sort of 'event' might have served equally well: a crime scare, a rumor of kidnapping by aliens, whatever."[25] Doubtless, Buell overstates the case; nonetheless, there is a metaphoricity to the event that renders it, much like the simulations practiced by DeLillo's Advanced Disaster Management company, an "all-purpose leak or spill. It could be radioactive steam, chemical cloudlets, a haze of unknown origin."[26] The airborne toxic event is, in other words, a form into which different content might be read; its very fictionality renders it mobile and flexible as allegory or metaphor. Indeed, if the event is anything, it is, as Frederick Buell points out, "a textbook case of Ulrich Beck's risk society."[27]

Given this openness, *White Noise*'s toxic event is perpetually timely, as the hazard du jour can be plugged into its metaphorical frame. Writing in the context of the first nuclear age, Michael Messmer argued, for example, that *White Noise* was particularly illustrative of a "nuclear culture," dominated by the "fabulous" simulacra of nuclear deterrence narratives. Framed by a description of a documentary about the Bikini evacuation, Messmer's article makes its own gesture toward "Bikini is our world" when he suggests that such theorists of the simulacrum as Jean Baudrillard, Umberto Eco, and, at least implicitly, DeLillo, "can help to implement thinking about the plight of the Bikini islanders and more generally about the nuclear culture in which all of us presently live."[28] In so doing, Messmer himself performs something like what he diagnoses, as the real of the plight of the Bikinese and the fake of simulation blur. The "fabulous textuality" of deterrence is the ultimate in simulation, for Baudrillard and Messmer alike, even as the absolute materiality of the experience of the Bikini islanders gets lost in the mythos of a simulacral culture. Messmer notes the similarities in DeLillo's description of the sublime airborne toxic event and the descriptions by those who witnessed the Trinity explosion, but the explosion in the novel is importantly not a nuclear fireball. Indeed, in his instructions to those participating in the "all purpose leak or spill" simulation, which, as I suggested, might be seen as a metafictional description of the novel's airborne toxic event, the SIMUVAC official explicitly says: "All you rescue personnel, remember this is not a blast simulation. Your victims are overcome but not traumatized. Save your tender loving care for the nuclear fireball in June."[29] A nuclear fireball may be an important thing to simulate, in other words, but for another novel.

Messmer is perhaps closer to an apt analogy when he turns to another potential referent for the event, the accident at the Three Mile Island nuclear power plant in March 1979, which, as he suggests, mobilized something like an "activist sublime," that "vague foreboding" that Babette feels in the face of the genetically engineered organisms. But considering the fact that the actual substance in the airborne toxic event is chemical rather than radiological, I would like to focus, instead, on another historical reference that has, in the intervening years, come to dominate DeLillo criticism, one that Messmer makes only parenthetically: Bhopal. This Bhopal gesture, a reference that rarely rises to the level of a full-blown comparison, was inaugurated by the early reviews and is fairly common in *White Noise* criticism, making Bhopal a kind of literary "sister city" to Jack Gladney's "Blacksmith." And though Bhopal is, clearly, just one among many potential historical referents, its privileged status arguably has been authorized by Mark Osteen's editorial decision to include a series of articles on Bhopal excerpted from *Newsweek* magazine as a component of the critical apparatus in his 1998 Viking edition of *White Noise*.[30]

At first glance, the idea that DeLillo's novel might offer insight into Bikini or Bhopal seems unlikely. *White Noise* could not be any less about Bhopal or the global village—or as Bradford would have it, Bhopal as the global village. The focus of the novel is quite myopic; DeLillo's characters consist of a narcissistic white middle-class family in a quaint college town, and though their lives are presumably shaped by a larger system that delivers their media access and flawless produce as well as the toxic cloud of pesticide by-product, Nyodene D, that dominates the second part of the book, this is a system that is largely invisible to them, as they visit the grocery store or watch television, as they participate in simulated disasters or go to the overpass to watch the glorious toxic sunsets. *White Noise* is many things:

a pointed satire of academia (the protagonist is the founder of "Hitler Studies"), a postmodern send-up (Baudrillardian simulacra abound), and an indictment of vapid consumerism. But, published in January 1985, *White Noise* cannot logically be a response to Bhopal, which occurred on December 3, 1984, at which point the novel was doubtless on its way to press, so any similarity in the novel's "airborne toxic event" to the Union Carbide explosion is coincidental. Nonetheless, as Osteen reports in his introduction to the critical edition, "When *White Noise* was first published in January 1985, reviewers were struck by its timeliness; indeed, appearing only a month after a toxic chemical leak at a Union Carbine plant in Bhopal, India, killed some 2500 people, DeLillo's novel—with an 'airborne toxic event' at its center—seemed almost eerily prescient."[31] Osteen does not assess the appropriateness of the comparison, and this absence of analysis leaves the question of the relationship of fictional and real "events" open to interpretation. But clearly such a gesture requires reducing Bhopal, metonymically, to the explosion of toxic gas and *White Noise*, synecdochally, to its "airborne toxic event," and to say that a novel about a white, upper-middle-class American family facing a toxic cloud with fairly nebulous health effects is prescient of an explosion in India that killed at minimum twenty-five hundred people instantly could be to say, in effect, "We all live in Bhopal."

If "eerie prescience" appears to suggest likeness in these events, however, this would seem to be refuted by comparing the actual staging of Bhopal in the *Newsweek* articles that Osteen includes to that of the event in the novel. DeLillo's characters flee the toxic cloud in their car; the effects of the chemical on them are nebulous. In Bhopal, on the other hand, as we learn in the articles, "It looked like a neutron bomb had struck. Buildings were undamaged. But humans and animals littered the low ground, turning hilly Bhopal into a city of corpses."[32] Whereas DeLillo's Jack Gladney has immediate access to the most cutting edge of medical technologies, the "most accurate test devices anywhere" with "sophisticated computers" that, in Jack's case, produce ominous "bracketed numbers with little stars"[33] indicating his Nyodene D exposure, in Bhopal, the *Newsweek* article tells us, "What the victims needed most, [according to] one harried doctor, were massive doses of antibiotics and vitamins"[34]—a diagnosis that points to the poverty present long before the explosion. Of course, given the fact that methyl isocyanate "belongs to a family of toxins for which there is no antidote and no treatment,"[35] vitamins and antibiotics seem as useful as the "high-potency antacid tablets" that the medical workers in Bhopal actually dispensed.[36] By the end of the articles, poverty itself seems to justify chemical exposure, as unnamed experts inform the reporters that "many of the victims in India would not be alive at all if not for chemicals that increased food supplies, reduced the incidence of malaria and improved chemical sanitation. Judged against such benefits," the authors conclude, "the risks of chemical accidents seem more acceptable."[37] Here, the "city of corpses" turns into an unfortunate but ultimately acceptable side effect, the collateral damage of an industrializing modernity.

Though such descriptions might seem to render Bhopal safely distanced, the kind of disaster that happens "over there" where disease and government corruption alike are rampant, however, the articles also speculate on the possibilities of such catastrophes occurring at home—playing with the likelihood that we all live in Bhopal by reintroducing the element of "chance" over "necessity." According to an expert on worker safety, "It's like a giant roulette game.... This time the marble came to a stop in a little place in India. But the next time it could be the United States," the stakes in that gamble seemingly raised by the relative health and wealth of the potential American victims.[38] Here, risk is a game of chance no longer rigged in the United States' favor. Such turns back to potential American risks suggest a narcissism in media staging that *White Noise* might be taken to satirize. In fairly nuanced version of the Bhopal gesture, in a January 1985 review of the novel, Jayne Anne Phillips highlights the distances between Bhopal and Blacksmith: "In light of the recent Union Carbide disaster in India that killed over 2,000 and injured thousands more, *White Noise* seems all the more timely and frightening—precisely because of its totally American concerns, its rendering of a particularly American numbness."[39] Like the novel's satirical professors of "American Environments," *White Noise* looks inward. Osteen's critical edition might offer an opportunity for counteracting this numbness, and thinking through the differences and similarities in the airborne toxic event and the Bhopal disaster is certainly potentially fruitful, making us aware of the noncontingent ways in which the accident is connected to its location. But the numbness to which Phillips refers might be not just to international events but to domestic concerns as well. Juxtaposing *White Noise* and Bhopal, whether for the purpose of remarking on prescience or diagnosing American myopia, may in fact collude in what Bradford calls the "deterrence machine," taking "our minds off the pervasive reality" that the novel represents, distancing the risk rather than bringing it home. After all, one need not dig into the nuclear archive or go to India to find a referent for a chemical explosion in the United States. As DeLillo reported in an interview, such disasters had been preoccupying him long before the catastrophic accident in Bhopal: "I began to notice something on television which I hadn't noticed before. This was the daily toxic spill—there was the news, the weather and the toxic spill. . . . This was one of the motivating forces of *White Noise*."[40]

REPATRIATING RISK

DeLillo's Jack Gladney does not live in Bhopal. But Union Carbide and its MIC gas are not exclusive to Bhopal either. As an experiment in risk repatriation, then, I would like to retain elements of the Bhopal gesture while situating it in the "American environments" that DeLillo quite explicitly describes. Doing so is in many ways against the spirit of the novel. As Michael Bérubé notes, "If we compare *White Noise*'s airborne toxic event with the 1984 toxic leak at the Union Carbide chemical plant in Bhopal, India (which occurred about a month before the book's publication and which DeLillo is often credited with having eerily anticipated), the motivelessness and agentlessness of [DeLillo's] event should seem all the more remarkable."[41] Bérubé's response is to read this motivelessness on its own terms (or, rather, in terms of *Underworld*), but I think filling in the gap in motive offers a way to remedy part of what I read *White Noise* as diagnosing. DeLillo's novel is, as Frederick Buell and Ursula Heise have noted, a satire of risk society, and though we could join his professors of "American Environments" in reading "nothing but cereal boxes,"[42] I submit that the novel's focus on commodified food (in the grocery store) and agricultural chemicals, specifically, suggests that tracking that cereal back to its ecological source might be more fruitful.[43]

Unmooring the novel from Bhopal renders Union Carbide and MIC also fairly arbitrary. Clearly, "Nyodene D," the chemical comprising the airborne toxic event, could represent any number of chemical toxins, and such chemicals were (and are) produced by a number of different corporations. Given that DeLillo's chemical is a by-product of the manufacture of insecticide—"The original stuff kills roaches, the byproducts kill everything left over," as Jack's son Heinrich quips[44]—however, and given the degree of uncertainty regarding Nyodene D's action on the body or the environment, MIC does offer a useful analogue. In this context, a more likely candidate for Blacksmith's sister city might be not Bhopal but the home of its "sister plant" in Institute, West Virginia, a community that, according to the authors of *No Place to Run*, experienced numerous toxic spills prior to the Bhopal explosion: "EPA found from the company's own records that it had leaked MIC 28 times during the five years ending in 1984," and "Carbide later admitted 62 leaks of MIC."[45]

The question reporters might have asked, then, is not "Could it happen here?" but "To what extent is it already happening here?" Even if we don't all live in Bhopal, we all may live near a Union Carbide (or now Dow Chemical) plant, and thus are also at risk. Indeed, subsequent legislation linked risk in India and the United States, suggesting that both the catastrophic accident and "a serious chemical release at a sister plant in West Virginia" together inspired the EPA's TRI (Toxics Release Inventory) Program. As Sheila Jasanoff notes, this phrasing "reduces to practically nothing the normative distance between a catastrophic industrial disaster in India and a routine chemical release in the United States."[46] Regulation of safety was wildly different for these two sister plants, a double standard that, for example, equipped both with scrubbers, but only one with the funding and personnel to maintain them. But even as we would clearly want to continue to emphasize the normative distance to which Jasanoff refers, we surely also do not want to dismiss the real hazard presented in the "routine chemical release," a phrase that sounds suspiciously benign.

The focus on Bhopal may, in other words, obscure other events that also chanced to coincide with the publication of the novel. Indeed, in *Advocacy after Bhopal* (2001), Kim Fortun describes an incident resembling the evacuation of the grade school in *White Noise*, as a result of the children "getting headaches and eye irritations, tasting metal in their mouths."[47]

> In November 1984, just two weeks before the gas leak in Bhopal, a "raw material" used to produce the pesticide Furadan spilled at an FMC plant in Middleport, New York. "Vapors from the spill entered the ventilation system of a nearby grammar school, requiring the evacuation of 500 children and the hospitalization of 9 of them, along with 2 teachers, due to eye irritation and respiratory difficulty. Later the city fire chief complained about the lapse of time between the actual spill and the notification of outside support agencies, school authorities, ambulance, fire and evacuation personnel." Later still Middleport residents found out that the chemical was MIC.[48]

In this case, DeLillo need not have been prescient, just descriptive. Such incidents give the lie to the comforting assurance that "industrial accidents have tended to occur in the Third World, where population density is higher and safety measures often fail to keep up with the spread of technology," a containment strategy that belies the actual failures to keep MIC and other dangerous chemicals contained.

Of course, these routine releases—or, to use DeLillo's phrase, "toxic spills"—also do not happen uniformly. Blacksmith comes under chemical cloud because these toxins are passing through en route to someplace else. Institute is, as the literature describing its vulnerability to contamination often emphasizes, a predominantly African American community. And, as critics have pointed out, DeLillo's Jack Gladney is conscious of the fact that he is generally the beneficiary of the uneven distribution of risk. As Jack puts it, "These things happen to poor people who live in exposed areas. Society is set up in such a way that it's the poor and the uneducated who suffer the main impact of natural and man-made disasters.... I'm a college professor. Did you ever see a college professor rowing a boat down his own street in one of those TV floods?"[49] The irony, of course, as Heise and others have noted, is that Jack asks this rhetorical question as he is himself fleeing an airborne toxic event, suggesting that even college professors are not immune.

Critics interested in the novel's Baudrillardian simulacra—or rather, implicit critique thereof—often point to the characters' dismay over the failure of the media to televise the airborne toxic event as yet more evidence of their divorce from the real, their inability to experience the material world around them in an unmediated fashion. But reading *White Noise* as about the risks that, say, a Union Carbide (or a Dow or a Monsanto) poses in the United States offers a somewhat different perspective. In a version of Beck's own suggestion that *"the more people who are poisoned, the less the poisoning takes place,"*[50] DeLillo's Babette wonders how serious a chemical cloud is in a context in which the toxic spill has become "routine": "Every day on the news there's another toxic spill. Cancerous solvents from storage tanks, arsenic from smokestacks, radioactive water from power plants. How serious can it be if it happens all the time?"[51] Despite the drama of their own "airborne toxic event," a catastrophe that requires mass evacuation and subsequent monitoring by men in "Mylex suits," it does not make the media. At one point, in the impromptu public sphere gathered at the evacuation center, an unnamed character reflects on the incongruity of this media silence:

> "There's nothing on the network," he said to us. "Not a word, not a picture. On the Glassboro channel we rate fifty-two words by actual count. No film footage, no live report. Does this kind of thing happen so often that nobody cares anymore? ... Are they so bored by spills and contaminations and wastes? Do they think this is just television? 'There's too much television already—why show more?' Don't they know it's real? Shouldn't the streets be crawling with

cameramen and soundmen and reporters? Shouldn't we be yelling out the window at them, 'Leave us alone, we've been through enough, get out of here with your vile instruments of intrusion.' Do they have to have two hundred dead, rare disaster footage, before they come flocking to a given site in their helicopters and network limos?"[52]

One could certainly read this, as John Duvall does, as yet more evidence that "DeLillo's postmoderns seek affirmation through television," which in turn "creates the Real."[53] Unable to experience the event in an unmediated fashion, the "tv man," as Duvall describes him, must see it externalized and rendered as spectacle. But this is potentially also something more than postmodern narcissism. The fact that toxic spills happen even in places like Blacksmith might, ideally, lead to something like a grassroots political consciousness, as characters with the financial and cultural capital to effect change are suddenly rowing that same boat down their streets, as it were. Instead, DeLillo stages the lack of staging, leaving the Gladneys and their neighbors isolated from the media public sphere. The comforting narrative about risk—that "these things happen to poor people who live in exposed areas"—remains intact, even potentially for the characters themselves.

White Noise thus raises doubts about the political efficacy of those aspects of contemporary culture that Ulrich Beck, in his dialectic of irony, hopes might generate a subpolitical response in risk society, the media and consumerism. As Duvall rightly points out, considering the Gladneys' relish in watching disasters on television, it is doubtful that televising their experience would have much impact, for any political consciousness seems foreclosed: "*White Noise* repeatedly illustrates that within the aestheticized space of television and the supermarket, all potentially political consciousness—whether recognition of the ecological damage created by mass consumption or an acknowledgment of one's individual death—vanished in formalism, the contemplation of pleasing structural features."[54]

Similarly emphasizing form over content, and commenting on the ways in which media serve to distance DeLillo's characters from the reality of what they are watching, Messmer cites this passage: "Only a catastrophe gets our attention. We want them, we need them, we depend on them. As long as they happen somewhere else. This is where California comes in. Mud slides, brush fires, coastal erosion, earthquakes, mass killings, et cetera. We can relax and enjoy these disasters because in our hearts we feel that California deserves whatever it gets." Messmer continues: "Suppose for the first 'California' we substitute 'Chernobyl,' for the second we substitute 'Russians,' and for 'lifestyle' we substitute 'Communist revolution.' Or this series of substitutions: 'Hiroshima,' 'Japan,' 'Japanese,' and 'Greater East Asia Co-Prosperity Sphere.'"[55] Messmer's point in these substitutions is that the structure of "the culture of simulacra and hyperreality," the culture that arises, at least in part, as a result of "nuclear culture," renders the content the same; be it California or Chernobyl, the Gladney family will watch with the same bemused detachment. One is tempted to repeat the experiment, inserting the Bhopal example, replacing "California" with "Bhopal" and "Indians," and for "lifestyle" substituting "poverty" or "government corruption." Certainly, the comforting notion that those who died in the Bhopal explosion would have died anyway, from poverty or disease, seems to fit the model.

But what is strikingly different about DeLillo's original choices and the places and terms that Messmer slots in is that California is not really "somewhere else" in the sense that Russia, Japan, or India is. The irony in the original passage is in the severe myopia that sees even California and its disasters as comfortably distanced from Blacksmith. In fact, despite the fact that, as Jack Gladney suggests of California, "we know we're not missing anything. The cameras are right there. They're standing by. Nothing terrible escapes their scrutiny,"[56] the characters do seem to miss the televising of the movement in California that might have helped them to connect the dots between their consumerism—especially at the supermarket—and toxic spills: the grape and lettuce boycotts spearheaded by the United Farm Workers (UFW), a movement to which, according to Ilan Stavans, "the nation tuned in."[57] What is striking about those boycott campaigns is the degree to which pesticide dangers offered a way to bridge the experiences of farmworkers and consumers, appealing not just to consumers' sense of justice but also to their fears of personal contamination. In campaigning for "votes," in the sense of the withholding of consumer dollars, César Chávez, the charismatic leader, who, Stavans reports, was "constantly in the spotlight, with cameras and microphones pointed at him,"[58] repeatedly invoked the universality of risk: "There is nothing we share more deeply in common with the consumers of North America than the safety of the food all of us rely upon."[59]

The elements for the success of this sort of political campaign are, in effect, all there in *White Noise*. Jack's wife Babette, with her constant purchases of yogurt and wheat germ and her part-time job teaching a course titled Eating and Drinking: Basic Parameters, would seem an ideal target for the UFW boycott, a health-conscious consumer, considering the future of her blended family. But though Jack finds in the produce isles of the supermarket "six kinds of apples," forever "in season, sprayed, burnished and bright,"[60] with what they might be "sprayed" is never a question. For all the characters' fears about electromagnetic radiation, toxic spills, "the water, the air,"[61] the supermarket remains an inner sanctum, closed off from the toxicity that is otherwise pervasive. Commenting on the potential occult meanings contained in the supermarket, Jack Gladney's colleague Murray suggests: "Everything is concealed in symbolism, hidden by veils of mystery and layers of cultural material. But it is psychic data, absolutely. . . . It is just a question of deciphering, rearranging, peeling off the layers of unspeakability. Not that we would want to, not that any useful purpose would be served."[62] This metafictional reverse psychology invites the reader to question what sort of defense mechanism underlies the refusal to "peel off the layers of unspeakability."

In the service of such peeling, one might juxtapose DeLillo's novel to Helena María Viramontes's *Under the Feet of Jesus*, a novel explicitly dedicated to César Chávez that highlights the experiences of farmworkers and their families. In a moment likely to be defamiliarizing for the average consumer, Viramontes offers an encounter with the reading of packaging very different from that performed by DeLillo's Murray. Here, a young farmworker discovers a distorted image of herself on what is clearly a box of Sunmaid raisins:

> Carrying the full basket to the paper was not like the picture on the red raisin boxes Estrella saw in the markets, not like the woman wearing the fluffy bonnet, holding out the grapes with her smiling, ruby lips, the sun a flat orange behind her. The sun was white and it made Estrella's eyes sting like an onion, and the baskets of grapes resisted her muscles, pulling their magnetic weight back to the earth. The woman with the red bonnet did not know this.[63]

If the woman on the box does not know this, the reader/consumer likely does not know it either, and the novel's defamiliarizing perspective works to render the unspeakable speakable, an approach all the more visceral when Viramontes describes the accidental poisoning of another farmworker who happens to be caught in the path of pesticides:

> Alejo's head spun and he shut his stinging eyes tighter to regain balance. But a hole ripped in his stomach like a match to paper, spreading into a deeper and bigger black hole that wanted to swallow him completely. He knew he would vomit. His clothes were dampened

through, then the sheet of his skin absorbed the chemical and his whole body began to cramp from the shrinking pull of his skin squeezing against the bones.[64]

Taking us behind the label, Viramontes reveals the "unspeakable," clearly in the hopes—as in the UFW boycott—that a "useful purpose would be served." Such representation may, to borrow from Upton Sinclair's account of the reception of *The Jungle*, "aim at the public's heart," a strategy certainly also used by the UFW, but it may also, as César Chávez did, work to "hit it in the stomach." To Jack's consumer observation of fruit, seemingly cut off from its origins in the fields, "gleaming and wet, hard-edged . . . like a four-color fruit in a guide to photography,"[65] Chávez would have noted that "innocent looking grapes on the table may disguise poisonous residues hidden deep inside where washing cannot reach."[66] Though pesticide exposures may seem limited to farmworkers—or to those "poor people who lived in exposed areas" that Jack describes—Chávez highlighted the persistence and potential universality of pesticide exposure, even for consumers who are far distant from the fields in question.

Publicizing these risks in the media was to lead to a consumer boycott, which would, in turn, transform the conditions under which farmworkers labored and fruit was grown, but this message never reaches DeLillo's Blacksmith. Indeed, in a kind of parody of boycott politics, Jack Gladney describes a trip to the mall in which his selective consumerism is entirely emptied of political implication: "We moved from store to store, rejecting not only certain items in certain departments, not only entire departments, but whole stores, mammoth corporations that did not strike our fancy for one reason or another."[67] Any connection between the actual actions of "mammoth corporations" and one's choice to buy or not buy is utterly effaced, producing the form of the boycott without any content. But Babette's observation about the prevalence of toxic spills may give some insight into the limitations of using consumer fear for political purposes. Reporting on the UFW boycott for the *New York Times* in 1988, David Wilson noted the increase in grape sales, despite widespread media attention: "Even some who support the boycott say the union has chosen a futile cause when consumers are bombarded with warnings about the radon in their cellars and the cholesterol in their eggs. 'People are getting numb to the issue of chemicals in the environment,' said Ron Pembleton, a director of the non-profit California Public Interest Research Group."[68] And certainly, as Ursula Heise has convincingly argued, toxic substances are pervasive in *White Noise*, whether in the accidental spills, odors, or side effects, or in the characters' intentional consumption of artificial sweeteners and pharmaceuticals.

There is clearly something catachrestic in reading DeLillo's referent as Bhopal—to do so requires that both DeLillo's imagined event and the disaster in Bhopal be extracted from their contexts, rendered as isomorphic figures of catastrophe, with all of the attendant political problems such reduction necessarily entails. But pursued fully, the Bhopal gesture also offers something useful to thinking global risk, as affixing the referents implied in the comparison—not only those in Bhopal, whose experience is decidedly unlike that in Blacksmith, but Union Carbide, MIC, Institute, Middleport, and those farmworkers in the United States who, as the authors of *No Place to Run* remind us, "are among the lowest paid and least protected of workers in [developed] nations"[69]—might render the novel's "symbols" legible in a way that "disclose[s] the structural character of the problems while at the same time fostering the ability to act," even if DeLillo's characters never quite read it in this way themselves.

LIVING IN BHOPAL

Who does live in Bhopal? How might the metaphor of "Bhopal" be mobilized in such a way as to emphasize both the similarities in the difference and the difference in the similarities? And, given that the Bhopal disaster is ongoing—the factory site still toxic, the groundwater poisoned, the injuries, to those present or born to those present, still untreated and potentially untreatable—how might we keep Bhopal in the public imagination, especially given the short attention span that the media clearly foster? Though the Bhopal disaster continues to make international news—from the controversial stunt in 2004, in which, claiming to be a Dow representative, one of the members of the activist group the Yes Men offered an apology and settlement for the disaster, to the more recent 2011 documentary film *Bhopali*, which chronicles the ongoing nature of the disaster—for literary critics and the general reading public alike, as Sheila Jasanoff reports, in "recent years, a work of fiction, Indra Sinha's novel *Animal's People*, may have done more to revive international interest in Bhopal, and thus to touch the conscience of the world, than decades of medical or legal action."[70] Set in the fictional Indian town of Khaufpur, *Animal's People* presents a thinly veiled representation of Bhopal some nineteen years after the night of the explosion, chronicling the ongoing disasters, environmental, health, social, political, and legal, that persist.

Reading *White Noise* as prescient of Bhopal produces its own sort of ominous irony. In the course of his musings on the televising of disaster footage, Jack Gladney's colleague Murray suggests that "India remains largely untapped. They have tremendous potential with their famines, monsoons, religious strife, train wrecks, boat sinkings, et cetera. But their disasters tend to go unrecorded. Three lines in the newspaper. No film footage, no satellite hookup."[71] By the time *Animal's People* was published in 2007, however, Sinha was able convincingly to portray his title character as having the relationship with the media for which DeLillo's "t.v. man" yearns ("Shouldn't we be yelling out the window at them, 'Leave us alone, we've been through enough, get out of here with your vile your vile instruments of intrusion'").[72] Of course, in this case, unlike the more "routine chemical release" depicted in DeLillo's novel, they indeed do have far more than "two hundred dead," plus "rare disaster footage." Sinha's novel, in staging for a world market the ongoing catastrophe of Bhopal, risks itself becoming so much sensational spectacle, offering up yet more "disaster footage" for a global market already saturated with catastrophe, a possibility to which the novel itself alludes. The premise of *Animal's People* is that it is a transcript of a series of tapes that a "native informant," Animal, a Khaufpuri resident who had been a child at the time of the explosion, and whose body is, as a consequence bent forward such that he walks on all fours, has agreed to make for a "jarnalis" (journalist) in exchange for a pair of shorts. As if imagining consumers like DeLillo's Gladneys, tuned in to their television to watch the disaster footage they so crave, Animal tells the journalist, "You were like all the others, come to suck our stories from us, so strangers in far off countries can marvel there's so much pain in the world."[73] The challenge of the novel is how to represent "so much pain," what Rob Nixon calls "slow violence," thereby encouraging the kind of cosmopolitan feeling Beck associates with global subpolitics, without allowing for the aesthetic distancing of the global consumer.

Nixon's essay (and, subsequently, chapter) on *Animal's People* is no doubt part of the reason that this novel has become, as Lawrence Buell has recently suggested, a kind of paradigmatic text for contemporary ecocriticism, replacing, in Buell's estimation, Rachel Carson's *Silent Spring* as representative of contemporary toxic discourse.[74] Whereas Carson constructed a fictional hamlet in her "Fable for Tomorrow" that offered, Buell asserts, a "pastoral-nostalgic memory of an idyllic middle-American town as ecoethical norm to counter the health hazards of chemical pesticides," Sinha refuses this sort of nostalgia. Indeed, at one point, Animal's adoptive "mother" figure, a nun whose explosion-induced aphasia has led to her inability to recall Hindi, recalls that, as a young child, Animal used to "enjoy swimming in the lakes behind the Kampani's [company's] factory. 'You'd dive right in, with your arms and your legs stretched out in one line.'"[75] But just a few pages later, Animal dispels any image of a bucolic past: "Just now I mentioned lakes, really they're clay pits behind the Kampani's factory where bulldozers would dump all different coloured sludges."[76] As several critics have noted, Animal's description of the former factory site, where abandoned chemicals await release, recalls Carson's "silent spring": "Listen, how quiet it's. No bird song."[77] And Nixon points out that one Khaufpuri response to the Kampani lawyers—"You were making poisons to kill insects, but you killed us instead"[78]—resonates with Carson's suggestion that "pesticide" and "herbicide" were misleading euphemisms that disguise the indiscriminate ways in which these "biocidal" chemicals destroy life. And yet, Nixon argues, "Sinha departs from Carson in representing 'pesticides' as both indiscriminate and discriminatory: their killing power exceeds their targeted task of eliminating troublesome insects, but they do discriminate in the unadvertised sense of saddling the local and global poor with the highest burden of risk. Thus, by implication, the biocidal assault on human life is unevenly universal."[79] In specifying the particularity of Khaufpur, *Animal's People* inevitably exposes the particularity of Carson's imagined town as well.

Juxtaposed to DeLillo's placid Blacksmith, Khaufpur could not be more different, from the poverty and garbage to the overt political activism, and clearly a comparison of the protagonists of these two novels highlights the disparity. Sinha's depiction of Bhopal could be read as a kind of postcolonial conscience for DeLillo critics, as, to borrow a distinction from Ramachandra Guha and Juan Martínez-Alier, the "full stomach" risk of the North confronts the "empty belly"[80] risk of the South, and Jack Gladney's nebulous "bracketed numbers with little stars" meet Animal's "coughing, frothing etc" followed later by "the smelting in [his] spine."[81] Whereas Jack's environs boast "the most accurate test devices anywhere," in Khaufpur, "all these years after that night," "there's still no real help for those whose eyes and lungs and wombs were fucked."[82] Whereas DeLillo's Jack speaks in first person to a general audience, thus offering the reader the comforting role of invisible confidant, Sinha's Animal is fairly explicit about the likely privileged status of his readers—whom he calls "eyes": "What can I say that they will understand? Have those thousands of eyes slept even one night in a place like this? Do those eyes shit on railway tracks?"[83] In other words, "we" don't all live in Bhopal; toxic as the planetary environment is, Bhopal is not (yet) "everywhere," as Bradford imagines.

For Animal and his "people," the boycott politics that Beck argues will provide the political engine of risk society, wherein dollars are votes that cosmopolitan consumers can use to influence corporations and governments, seems patently impossible. As the political activist Zafar, puts it, "We people have nothing, many of us haven't an untorn shirt to wear, many of us go hungry, we have no money for lawyer and PR, we have no influential friends."[84] Refusing to buy products produced by Union Carbide (or Dow) is clearly not an option. Whereas the "kampani" is mobile, able to capitalize on cheap labor and lax regulation only to depart when the inevitable results of those choices came to pass, these residents of world risk society are rooted firmly in toxic place. The consumerist choice exercised by DeLillo's Jack, "rejecting not only certain items in certain departments, not only entire departments, but whole stores, mammoth corporations that did not strike our fancy for one reason or another," is not available.

For these characters armed with the "power of nothing,"[85] what they can withhold is their bodies, and in this way the novel does stage a kind of boycott. When an American woman, Elli, comes to town to open a new free clinic, Sinha's Khaufpuris are understandably suspicious that the Kampani might be funding it, since the Kampani would have an interest in reading the data their bodies might produce in ways that shift responsibility off of itself. Zafar notes the coincidence of a major court victory for the victims of "that night" and the government's approval of the free clinic:

> "Think like the Kampani. Thousands of people say that for twenty years their health's been ruined by your poisons. How do you refute this? We say that the situation is not as bad as alleged, that not so many people are ill, that those who are ill are not so seriously ill, plus of whatever illnesses there are, most are caused by hunger and lack of hygiene, none can be traced back to that night or to your factory."[86]

In this case, the turning of bodies into data—Jack Gladney's "bracketed numbers and pulsing stars"—and the expert knowledge necessary for its deciphering, becomes just one more means to the end of abdicating corporate responsibility. Such paranoia is hardly unfounded, recalling *Newsweek*'s "harried doctor" who diagnosed "massive doses of antibiotics and vitamins" for the victims, or the "expert" who suggested that industrial accidents were a small price to pay for the laudable benefits of "chemicals that increased food supplies, reduced the incidence of malaria and improved chemical sanitation," without which "many of the victims in India would not be alive at all."

With no dollar/votes to withhold from the global economy, the residents of Khaufpur thus withhold their bodies, refusing to provide the data on which they suspect subsequent exonerations might be built. But this strategy ends up having been misdirected. The clinic is not, as it turns out, in league with the Kampani, and their bodily withholding does more to damage themselves than it does the invincible (and largely invisible) corporation. With the failures of this strategy, the activists turn to another form of bodily withholding, in this case, staging the wasting of their own bodies in a hunger strike, but as Zafar's girlfriend Nisha tells him, such staging depends on empathy, which the Kampani clearly does not have. As Zafar nears death, Nisha admonishes, "If there was one good person in that Kampani, even one who might be moved by such a sacrifice, then it might be worth it."[87]

As it turns out, the effective strategy for influencing Kampani decisions is not empathy but fear. When the Kampani lawyers appear in Khaufpur, staying at the expensive hotel outside of the city, their closed-door meeting with local politicians is interrupted when a mysterious woman clad in a black burqa (whom the novel implies may be Elli) empties a "bottle of stink bomb juice" into the air conditioning system.[88] Animal describes the scene: "These Kampani heroes, these politicians, they were shitting themselves, they thought they were dying, they thought they'd been attacked with the same gas that leaked on that night, and every man there knew exactly how horrible were the deaths of those who breathed the Kampani's poisons."[89] This theatrical staging of catastrophe is akin to Beck's description of the function of an organization like Greenpeace, whose agents he describes as "multinational media professionals who know how self-contradictions between pronouncements and violations of safety and surveillance norms can be presented so that the great and powerful (corporations, governments), blinded by power, stumble into the trap and thrash around telegenetically for the entertainment of the global public."[90] In *Animal's People*, similarly, "What made the whole thing fully grand was that someone had tipped off the press, they were waiting with their cameras when these goons stumbled out into the lobby."[91] Their secret backdoor machinations exposed, the corporate and government representatives are no longer able to avoid the public forum of the courts. Animal rejoices in the irony of subjecting the Kampani to its own poison, at least metaphorically, suggesting that the stunt produces "poetic justice of the fully rhyming kind." But Zafar reminds him that "poetic justice, rhyming or not is not the same as real justice."[92] Nevertheless he adds: "But being the only kind available to the Khaufpuris was at least better than nothing."[93]

This reference to a justice that is "poetic"—or at least representational—can only, in a novel that itself stages the ongoing catastrophe in Bhopal, read as metafictional. The Khaufpuris, with the power of nothing, are not in a position to "vote" with their dollars (or rupees), but Sinha's international audience probably is. At the beginning of the novel, as Sinha sets up the premise of Animal as native informant, he alludes to a letter from the journalist who gave Animal the tape recorder into which he is presently narrating his story. The letter reads: "Animal, you think books should change things. So do I. When you speak, forget me, forget everything, talk straight to the people who'll read your words. If you tell the truth from the heart, they will listen."[94] At once emphasizing and de-emphasizing his own mediation, the journalist/Sinha connects readers to the subaltern figure empathetically, our cosmopolitan feeling potentially leading to the "cross-border compassion" that Beck describes.[95] And certainly, as Sinha knows, the Bhopal disaster continues: Union Carbide (since absorbed by Dow Chemical) still evades responsibility; the site continues to be toxic; the people continue to suffer. Sinha knows firsthand the potential power of media staging to win dollars for causes. He describes himself as an "accidental activist."[96] A former copywriter for an advertising agency, Sinha was commissioned by a Bhopali activist to produce an ad that ran in the *Guardian* newspaper. Bron Sibree describes the results: "Accompanied by Ragu Rai's iconic photograph of a Bhopali child's burial, it convinced the public to donate £60 thousand—enough to build the Sambhavna Clinic and hire medical staff."[97] This backstory offers insight into the potential source of funding for Elli's free clinic, not the cynical machinations of corporate data-shaping, but the empathetic donations of a canny cosmopolitan audience. Still, as Zafar suggests, true justice requires more than this sort of philanthropic gesture; Sinha's own poetic contribution to the struggle offers a way to sustain attention beyond his ad campaign, a justice inadequate still, but better than nothing.

BHOPAL AS FIGURE

Sinha's Khaufpur is uncannily like Bhopal, from the language and culture to the particularities of the toxin, the corporation, and the political and legal wrangling that surround efforts to clean up the site and compensate the victims. (The clear impossibility of either of these tasks is in no way an excuse for abdicating responsibility for performing them, of course.) But Sinha insists that "Khaufpur shares things with, but is not, Bhopal."[98] Indeed, as Rob Nixon notes, quoting Sinha, "We can recognize Khaufpur as both specific and non-specific, a fictional stand-in for Bhopal, but also a synecdoche for a web of poisoned communities spread out across the global South: 'The book could have been set anywhere where the chemical industry has destroyed people's lives,' Sinha observes. 'I had considered calling the city Receio and setting it in Brazil. It could just as easily have been set in central or south America, west Africa or the Philippines.'"[99] If Sinha's observation suggests synecdoche, however, Nixon's own tropological wager is something closer to analogy, for he juxtaposes Sinha's fictional depiction of "biological citizenship" in Bhopal to Adriana Petryna's nonfictional account of "biological citizenship" in the aftermath of Chernobyl, using these two otherwise disparate examples—chemical versus radioactive toxins, postcolonial versus post-Soviet contexts—to demonstrate a continuity in the biopolitical consequences of "slow violence." Both Bhopal and Chernobyl were explosive disasters with long aftermaths, in both cases ongoing, and both involve what Nixon calls a "foreign burden," not just insofar as bodies carry toxins from the outside, but also insofar as the toxins are themselves "outsiders," whether originating from a U.S. multinational or a defunct Soviet Union.

This kind of analogical work—pointing out the ways in which two otherwise disparate disasters are nonetheless alike—offers fodder for a transnational activist imagination, a means by which a collaborative critical project that responds to a world at risk might be effected. And it is, again, encouraged by Sinha's *Animal's People* as well. In the midst of his hunger strike, when Zafar's attention would seem to be most focused on the particularities of his own body in the limited space of Khaufpur and the particular goal of forcing the Kampani to face the victims in court, he instead takes the opportunity to reflect upon his likely relationship to a series of analogously poisoned places: "Is Khaufpur the only poisoned city? It is not. There are others and each one . . . has its own Zafar. There'll be a Zafar in Mexico City and others in Hanoi and Manila and Halabja and there are Zafars of Minamata and Seveso, of Sao Paolo and Toulouse and I wonder if all those weary bastards are as fucked as I am."[100] Here, to the radiation at Chernobyl, Zafar adds the mercury contamination at Minamata and the dioxin at Seveso, all alike in being "poisoned cities," as disparate as each individual context clearly is.

In juxtaposing Petrya's anthropological account of Chernobyl with Sinha's fictional account of Bhopal, Nixon might seem to be comparing apples and oranges, a contrast heightened by the fact that he emphasizes genre in identifying *Animal's People* as the "environmental picaresque." But literature qua literature, the archive that produces and contains its own referent, necessarily leaks in the second nuclear age, producing just this sort of incongruous analogue. Here, too, Sinha offers a usefully metafictional reference. At one point, hoping to square the worldviews of Elli and Nisha's father, Somraj, Animal asks Zafar whether seeing the world as made of promises is compatible with seeing it as comprised of music. Zafar replies: "Likening music to promises is as absurd as comparing a vulture and a potato, potatoes don't have feathers and vultures don't grow under the earth. . . . You are making an equation of two things which have nothing in common."[101] When Animal asks, "What is an equation?" Zafar answers: "A way of showing how two different things can be the same."[102] But Animal rises to the representational challenge involved in rendering vultures analogous to potatoes by pointing out the coincidence in size in a potato and vulture's egg.

A potato and a vulture might be an apt figure too for the juxtaposition of *White Noise* and *Animal's People*, two texts that, apart from their generic coincidence, might seem to have little in common. But, in a kind of reversal of what I have called the Bhopal gesture in DeLillo criticism, Nixon cites *White Noise* in a footnote in which he suggests that DeLillo's novel, in which exposure to the airborne toxic event turns Jack into "the sum of [his] data," offers a "shift toward a different mode of biological citizenship."[103] The implication here is that whereas victims in Chernobyl and Bhopal seek restitution through medical representation, the victims in *White Noise* have representation without any such compensation—in this sense, Jack's experience is not wholly dissimilar from that feared by the residents of Khaufpur, who suspect that their bodies will provide data for a "massive database tally"[104] without any compensatory remediation of health or environment. Though Nixon does not pursue the DeLillo/Sinha comparison, his inclusion of it begins to suggest the series of fictional and nonfictional stagings that together might comprise an archival counter to the "database tally" of global risk.

"THE APOKALIS HAS BEGUN, AND THE WHOLE WORLD'S FULL OF IT"

Of course, even as we might see some poetic similarities in a potato and a vulture, some commonalities in diverse ecocatastrophes, such work of translation does as much to reinforce difference as it does to suggest similarity. "We all live in Bhopal" will only work as an activist strategy when it retains also the particularity of each analogical example. Here too we might take some inspiration from *Animal's People*. Animal's adoptive mother, Ma Franci, is a French nun who suffers from a particular form of aphasia following the explosion at the factory: "She'd gone to sleep knowing [Hindi] as well as any Khaufpuri, but was woken in the middle of the night by a wind full of poison and prophesying angels."[105] This aphasia leaves her not only monolingual (or at least so she believes; in fact there is some Hindi and English mixed in) but also incapable of recognizing other language as language; instead, she hears "stupid grunts and sounds."[106] This form of aphasia is one that Roman Jakobson, in his foundational essay "Two Aspects of Language and Two Types of Aphasic Disturbances," described as a "similarity disorder," one that results in " a confinement to a single dialectical variety of a single language" and a sense that others' utterances are "gibberish or at least in an unknown language."[107] This failure to translate from one language to another is related to a failure to discern similarity, and thus a failure of metaphor, as opposed to metonymy, which operates more by contiguity. Having been trained in the Christian tradition and having experienced the Khaufpur disaster, Ma believes that the apocalypse has literally come, the contingency of story and event has become necessity, and thus "The Apokalis has begun, and the whole world's full of it."[108] The global reach of Ma's phrase is akin to "We all live in Bhopal"—and thus is, in effect, metaphorical—but she is unable to read it as metaphor, which requires, as David Lodge reminds us, a feeling of disparity in vehicle and tenor.[109] As Ma Franci carries her literal reading further, turning gas victims into angels, Animal despairs, "wondering how anyone can get it so totally wrong," but despite her inability to translate among different semantic levels, the nun may, in fact, not be "totally wrong."

The "whole world" may not be literally in Khaufpur, but insofar as Khaufpur, like Bhopal, is a "poisoned city" among others, the whole world may be characterized by the slow apocalypse. Indeed, one of Sinha's most colorful characters, Kha in a Jar, a two-headed fetal victim of "that night," informs Animal that "everyone on this earth has in their body a share of the Kampani's poisons."[110] This, too, might be taken as metaphor, but reading Sinha's "Kampani" as Union Carbide suggests something more literal. On the one hand, my own opening juxtaposition of Bikini and Bhopal might appear as so many apples and oranges or vultures and potatoes, a kind of parataxis of similar rhetorical strategies intended to bring toxic contamination home, to suggest that these seemingly isolated catastrophes are part of a larger systemic catastrophe. On the other hand, in the examples of nuclear fallout and methyl isocyanate the similarities are also more than merely an accident of rhetorical figure. Union Carbide had been, after all, long involved in the nuclear industry, processing uranium and plutonium at facilities in Kentucky and New York. When, in *Animal's People*, Zafar has a dream in which he sees the Kampani as a giant edifice whose "basements contain bunkers full of atomic bombs," this is a revelation of historical fact.[111] And when, anticipating the dangers of lax safety regulation at the Bhopal plant, an engineer said, "They're putting an atomic bomb in the middle of your factory that could explode at any time," his seeming metaphor offered more than a powerfully affective rhetorical flourish.[112]

Union Carbide's pesticide-manufacturing facilities in India were, of course, aimed, not at arming a nuclear weapon, but at defusing another "bomb," the term Paul Ehrlich infamously mobilized as a metaphor for global population.[113] Indeed, in conceiving this global risk, Ehrlich employed a strategy structurally similar to Bradford's and Bradley's, figuring a particular place as the emblem of the disaster secretly happening on an otherwise inconceivably global scale. In the opening pages of *The Population Bomb* (1968), Ehrlich describes feeling affectively the problem of population first on a visit to India: "The streets seemed alive with people," he recalls. "People eating, people washing, people sleeping. People visiting, arguing, and screaming. . . . People, people, people, people."[114] This experience induces a moment of what might be called "population panic," as he and his less-than-replacement family (wife and one child) wondered, "Would we ever get to our hotel? All three of us were, frankly, frightened."[115] Like the literal "bomb," this scene with its "hellish aspect" was intended to deter, to prevent the "bomb" from falling on us all, but it was also intended to provide a kind of secret expression of what the world already was becoming, as, in an inversion of atomic risk, the peril—and responsibility—shifted from the destruction of life to the problems of its flourishing.[116]

Perceiving some "secret" collusion in the coincidence of Union Carbide's nuclear and pesticide activities might seem so much conspiracy theory, but, as Susan Mizruchi argues, attention to risk requires the critic to be "paranoid as well as creative,"[117] and both nuclear fallout and MIC expose the thanatopolitical underside of the biopolitical triumph of modernity.[118] If the bomb might appear to be clearly on the side of death and the green revolution on the side of life, if pesticides, herbicides, industrial fertilizers were precisely those technologies that would "make live" those populations that otherwise seemed themselves a kind of "bomb" on the earth, this is perhaps because the death associated with agricultural and other chemicals is precisely on the order of collateral damage—the kind of devastation central to atomic testing, as those in the Pacific, in Nevada, in the Aleutians, and elsewhere could attest, but that was often unseen in the face of the "fabulous" war to come. As farmworker organizer César Chávez said of chemical pesticide producers, "They would have us believe they are the health givers—that because of them people are not dying of malaria and starvation."[119] Highlighting the violence that underlies this rhetoric, Chávez continues, "It's a lot like that saying from the Vietnam War: we had to destroy the village in order to save it. They have to poison us in order to save us."[120] If, according to Foucault, biopower consists, no longer of the power to "take life or let live," but of the power to "'make' live and 'let' die,"[121] both atomic testing and agricultural chemicals operate in this latter fashion: In order that some might live, some must be sacrificed, whether as Bikinese or Bhopalis; downwinders or farmworkers. To say "We all live in Bhopal" is thus to declare solidarity with those sacrificed to modernity, but it is also, as activists from Bradley to Bradford to Chávez suggested, a revelation of self-interest. Even Rob Nixon, whose book repeatedly reminds us about the particularities of risk, the unevenly biocidal effects of toxins, warns, "We're all downwinders now," though "some sooner than others."[122]

Writing in 1986, Shiv Visvanathan reported: "Bhopal is still a catastrophe in search of a metaphor, a vision that is more and less than Bhopal."[123] The Bhopal gesture in DeLillo criticism suggests the extent to which Bhopal as a catastrophe has become a metaphor. Positioned as the revelation of the real to which *White Noise* was unknowingly referring, the catastrophe about which the novel was "prescient," Bhopal becomes mobile, flexible, accommodating the "littler Bhopals in our lives and yet . . . fluid enough to sensitize us to the destructive aspects of science and industrialism."[124] Such mobility risks, of course, collapsing that "normative distance between a catastrophic industrial disaster in India and a routine chemical release in the United States" that critics like Jasanoff want to emphasize. There was, most assuredly, a double standard in the Union Carbide–India and Union Carbide–U.S. "sister plants" that must not be effaced. Nonetheless, mobilizing "Bhopal" as metaphor helps to keep it in the public imagination, supplying an ongoing symbol to "touch and alarm" the "cultural nerve fibres" of world risk society.

THREE | # Discomfort Food: Analogy and Biotechnology

Genetic engineering is to traditional crossbreeding what the nuclear bomb was to the sword.
 —Andrew Kimbrell[1]

There has been no biotech Chernobyl.
 —Kerry Whiteside[2]

Okay, deep breath—it just isn't right to criticize genetic engineering as unnatural, as if decent people should ban horses, dogs and cats, wheat and barley. . . . What's wrong with genetic engineering is that it turns life forms into private property to enrich huge corporations.
 —Timothy Morton[3]

I see a close analogy between John von Neumann's blinkered vision of computers as large centralized facilities and the public perception of genetic engineering today as an activity of large pharmaceutical and agribusiness corporations such as Monsanto. The public distrusts Monsanto because Monsanto likes to put genes for poisonous pesticides into food crops, just as we distrusted von Neumann because he liked to use his computer for designing hydrogen bombs secretly at midnight.
 —Freeman Dyson[4]

TROPOLOGICAL WAGER: ANALOGY

What's wrong with genetically modified foods—those products judged "substantially equivalent" and fed to North American consumers willy-nilly;[5] the same ones rejected for so long (and in some quarters still) by the European Union, and turned back at one point by the boatload when sent to Zambia or, more recently, China[6]—the "StarLink™" corn, the "Roundup Ready®" canola, the "Genuity™" soybeans? What's wrong with these foods, and, perhaps even more pointedly, how do we know? Are transgenic foods like nuclear weapons (as the epigraph above that I have drawn from Andrew Kimbrell seems to suggest)? Will there be, as Kerry Whiteside anticipates, a "biotech Chernobyl"? Or are GM foods a benign solution to an intractable problem, a way to reduce chemical inputs, thus avoiding the catastrophes of pesticide manufacture of the sort described in *White Noise* or *Animal's People*? Are genetically modified foods truly analogous, as ecocritic Timothy Morton has asserted, to "horses, dogs, cats, and barley"? If access to genetic engineering were, as the physicist Freeman Dyson has recently imagined, democratized, modeled on open-access software—available to anyone to "download" as "do-it-yourself kits for gardeners, who will use genetic engineering to breed new varieties of roses and orchids. Also kits for lovers of pigeons and parrots and lizards and snakes to breed new varieties of pets"[7]—would the objections to genetic engineering be rendered moot?

If the previous chapters looked backward, to origins and past catastrophes, the present chapter looks forward, toward potential catastrophes that may—or may not—be in the process of becoming real. What genetically modified foods will have been will have implications on a global scale, for these organisms are planted on every continent, and, in principle, are self-replicating. And, as the flurry of questions with which I began suggests, the degree to which this will have been benign or malign is yet to be seen. If the examples of Bikini and Bhopal, or Bhopal and Chernobyl directed attention to the potential for the uses of analogy in linking seemingly disparate catastrophes, the example of genetically modified organisms asks that we extend the use of analogy in the service less of diagnosis than of prognostication. And here the protocols of precautionary reading become as important as they are freighted with additional risks. My focus here is the discomfort—critical, ethical, political—that surrounds genetically modified foods, and in my opening question I have no doubt tipped my hand, for this discomfort is in part my own. In feeling it, I am hardly alone; indeed, what could be more "natural" than to express such discomfort, one shared by people from such diverse walks of life as activist-scholar Vandana Shiva, Venezuelan president Hugo Chávez, Pope John Paul II, and Britain's Prince Charles?[8] (But now, surely, I have moved from describing discomfort to eliciting some of it on the part of my readers, for opposition to GM foods can produce some distinctly uncomfortable alliances.) Is genetic modification just the latest in a long process that began with domestication and moved through hybridization (an argument often made to alleviate anxiety), or is it a dramatic leap into a new process altogether (an argument that has accompanied patent claims)? Will genetically modified crops feed the planet's hungry or prove carcinogenic—or both?[9] Will it have mattered that the bacterial insecticide Bt was not only on but *in* our soy?

To these questions, I must respond, at the outset, that I do not know. Of course, lack of expertise in biology has not stopped literary and cultural critics from commenting on genetic technologies in the past. Indeed, whether it is because we cannot resist intervening in discourse surrounding something persistently referred to as "like a language," or simply because we, like all human and nonhuman beings on the planet, are increasingly surrounded by new life-forms—including transgenics, which cross unrelated species, phyla, and even kingdoms, joining fish and tobacco, bacteria and trees[10]—that fascinate us, literary and cultural critics have produced a fairly substantial archive of responses to genetics in general and genetic engineering more specifically, our relative lack of expertise notwithstanding.[11] Genetically modified organisms clearly put literary and cultural critics solidly in the position of being *"dependent on second-hand non-experience controlled by professionals outside their field,* with all the damage that does to their battered ideals of professional autonomy."[12] In our "not-knowing," we are, however, in good company, for as Ulrich Beck points out, when it comes to the potential hazards of new technologies, "No one is an expert—especially not the experts."[13] Even those expert scientists with *firsthand experience* (as opposed to "second-hand non-experience") of GM crops are not in agreement on the subject. Genetically engineered food is, thus, as Claire Hope Cummings puts it, an "uncertain peril," in part because it is the peril itself that remains uncertain.[14]

Of course, my opening question—what's *wrong* with genetically modified foods?—might itself cause some critical discomfort, for it implies a moral register all too common in these debates, a register that literature is often called upon precisely to provide. And even if risk critics cannot claim any specialized knowledge of gene transfer, we do have a sort of expertise in the structure of feeling that might accompany such technology, for literature has long been a locus for expressing doubts about the hubris of science. From Nathaniel Hawthorne's "The Birth-Mark" to Mary Shelley's *Frankenstein* to Aldous Huxley's *Brave New World*, works of literature have arguably anticipated Beck's risk society by centuries, even if the risk—and what is perceived to be at risk—changes over time. It is hardly surprising, then, that allusions to these texts often appear in the debates on biotechnology, as, for example, in former president Bush's Council on Bioethics, which took "The Birth-Mark" as a primary text in considering the ethics of cloning.[15] Indeed, when microbiologist and biotech advocate Walt Ream titled his pro-GMO piece published in a 2009 issue of *Microbial Biotechnol-*

ogy "Genetically Engineered Plants: Greener Than You Think," presumably tapping into the "ecofriendly" connotations of "green," he unwittingly (and ironically) alluded to an apocalyptic novel by Ward Moore, *Greener Than You Think* (1947), in which a genetic technology intended to facilitate a global green revolution ("no more famines in India or China") ends up enhancing suburban lawns, which then turn into a monstrous "GRASS MENACE"[16] the progress of which even the military is powerless to halt.

The lesson in these literary texts seems to be something along the lines of "Don't tamper with Mother Nature," a model often viewed as problematically conservative or even reactionary. Preferring not to get embroiled in such morally fraught positions, cultural critics have generally approached GMOs through what in bioethics debates are called "extrinsic" concerns rather than "intrinsic" objections—a distinction usefully summarized by Bernard Rollin, who explains the latter as a belief that "the wrongness of the action is not alleged to be a function of pernicious results or negative utility or danger—it is just wrong."[17] I invoke the moral register ("wrongness"), here, less to determine what the appropriate response to genetic modification is than to explore the relationship of both intrinsic and extrinsic responses to the larger context of risk society. When we do not know what the "pernicious results" of technology might be, moral discourse often substitutes for "not-knowing," offering a telling symptomatography, not only for other cultural concerns with which genetic engineering has become intertwined, but also for the gaps and absences in knowledge that pro-biotech discourses tend to suture over in the rush to market.

That critics of genetically modified foods have drawn on the rich metaphoric reservoir provided by mythology and literature is commonly acknowledged. From "Pandora's picnic basket"[18] to "Frankenfoods," such allusions have provided not only political keywords that tap into cultural histories of discomfort but also fodder for the counterargument—by those GM boosters who claim that objections to these novel organisms are unscientific, based in irrational fears. Given the uncertainty that surrounds these novel organisms, most attempts to answer the question of what might be wrong with genetically modified foods deploy the strategy of analogy, from Timothy Morton's homey "horses, dogs and cats, wheat and barley" to Kimbrell's ominous nuclear bomb and Whiteside's biotech Chernobyl. But deployment of such analogies is not limited to biotech's lay commentators. Indeed, the modes of knowledge production offered by novelists or literary critics and regulatory "experts" are not as far apart as some GMO supporters represent them to be, for the regulatory procedure that has, at least in North America, satisfied the authorities on the safety of GM foods, "substantial equivalence," too draws explicitly on the strategy of analogy. Introduced by OECD (Organization for Economic Co-operation and Development) in 1993, the doctrine of substantial equivalence suggests that "for foods and food components from organisms developed by the application of modern biotechnology, the most practical approach to the determination is to consider whether they are *substantially equivalent* to *analogous* food product[s]."[19] As in Morton's assertion that the results of genetic modification are analogous to "horses, cats and dogs, wheat and barley," the move, here, is, in effect, from simile—a GM food is *like* a non-GM food—to synecdoche—which conveniently dispenses with the implied nonequivalencies that are not included in this "substantial"-part-for-the-whole.

Of course, analogy is a fitting strategy for grappling with new technologies and the risks they may entail, for, as Eve Tavor Bannet suggests, analogy is "a method of reasoning from the known to the unknown."[20] But in the process, it builds in a nagging discomfort, perhaps particularly for those of us trained in literary studies; as Bannet notes, "For traditional rhetoricians, an analogy is not an identity; it is a figure which marks both the likeness and the difference in our application of words from case to case. The gaps, the discontinuities, and the differences are as important as the likenesses."[21] Reading "substantial equivalence" in this light is, in fact, not so willfully naive as it might seem, for it has been widely acknowledged that this analogical "test" is not based in science. As Erik Millstone, Eric Brunner, and Sue Mayer argued in the "Commentary" section of the journal *Nature* over ten years ago, "Substantial equivalence is a pseudo-scientific concept because it is a commercial and political judgement masquerading as if it were scientific. It is, moreover, inherently anti-scientific because it was created primarily to provide an excuse for not requiring biochemical or toxicological tests."[22] Many GMO proponents acknowledge these limitations, insisting that it is a conceptual tool, a "*starting* point of the assessment rather than an *end* point."[23] But of course the "starting point" for an analogy shapes its end. The vehicle of the conventional food lends itself, clearly, to the tenor of normalcy—a model that critics like Whiteside or Kimbrell attempt to overturn by reference to nuclear weapons or meltdowns.

RISK REALISM: STAGING UNCERTAINTY

The problem, of course, in analogies for genetically modified foods is that each sutures over the gap in knowledge that is basic to our present condition. Certainty about what GMOs will have been requires an extrapolative imagination—envisioning a future that, in official regulatory models, is very like our present (or, in boosters' views, much improved) and, in activist imaginaries, is often quite different. For these latter readers of GMOs, as the term "Frankenfoods" suggests, the genres most often associated with genetic modification are science or extrapolative fiction, the fantastic, even the gothic. And certainly future-oriented works like Margaret Atwood's recent *Oryx and Crake* (2003), *The Year of the Flood* (2009), and *MaddAdam* (2013) offer useful extrapolations of present practices. Her Pigoons, Rakunks, ChickieNobs, and the like offer dramatic images of Frankencreatures—not that far removed from our transgenic animals or cultured meat—that may indeed be on the horizon.

In those extrapolative imaginings, however, the questions raised by GM foods are largely answered: these creatures are indeed different from our "horses, dogs and cats, wheat and barley," not just in degree, but in kind. As with most extrapolative fiction, the point is, first, to suggest that the means to the apocalyptic futures are already in the works and, second, to prevent the outcome imagined ("if this goes on . . ."), or, as Ursula K. LeGuin said in her introduction to her own more properly science fictional text, *The Left Hand of Darkness*, speculative or extrapolative fictions "generally arrive about where the Club of Rome arrives: somewhere between the gradual extinction of human liberty and the total extinction of terrestrial life." LeGuin concludes: "Almost anything carried to its logical extreme becomes depressing, if not carcinogenic."[24] The rhetorical power of such narratives may certainly be deployed in the interest of anti-GMO sentiment, but, in carrying present realities to their extreme "carcinogenic" conclusions, they also potentially oversimplify our present condition, the "reality" within which we must live. Speculative fiction can have a precautionary effect, but only by forecasting a future that, in risk society, is necessarily unpredictable.

In order to explore "what's wrong with genetically modified foods," then, I turn in this chapter to an example of risk realism, a novel that grapples with the question of what it might mean to dwell in our present condition of uncertainty—and that, in that dwelling, confronts the issue of "wrongness" with all of its moralizing potential—Ruth Ozeki's *All Over Creation* (2003). Fictionalizing events surrounding the introduction and eventual withdrawal of the Monsanto corporation's "NewLeaf™" potato (thinly disguised in the novel as the "Cynaco" corporation's "NuLife" potato), Ozeki's novel is at once a depiction and itself an example of anti-GMO discourse that attempts to navigate the problem of how to represent a danger about which "no one is expert—especially not the experts." In this sense, *All Over Creation* is somewhat different from her earlier *My Year of Meats* (1998), a novel much more in the muckraker tradition, in which documentarian-protagonist Jane Takagi-Little exposes the legacies of DES (Diethylstilbestrol), a synthetic estrogen given to women to prevent miscarriage and to cows to promote growth. As Jane discovers, not only did DES not prevent miscarriage, it was carcinogenic and teratogenic, causing reproductive anomalies in children. Taking a job creating a glorified infomercial for the beef industry, Jane learns, not only that she is herself a DES daughter, but that DES and other synthetic hormones are still being used in beef production, likely causing the disruptions in endocrine systems that she observes in those working in the industry. Clearly itself an exposé, *My Year of Meats* leaves the reader feeling fairly certain of its stance on hormones in meat, and though her potato novel is similarly critical of unsustainable agricultural practices, it is, following the issue it traces, less concrete about what, exactly, ought to authorize that critique.

Both *My Year of Meats* and *All Over Creation* thus fall into the genre that Lawrence Buell has called "toxic discourse," those fictional and nonfictional texts (Rachel Carson's *Silent Spring*, Terry Tempest Williams's *Refuge*) that counter prevailing "expert" knowledge by asserting the dangers of chemical, nuclear, or other risks. Given that Buell's examples are chemical and radiological, something more along the lines of *My Year of Meats*'s DES, Ozeki's *All Over Creation* can be read as an extension of toxic discourse into new territory, for if, as Buell says of the hazards represented in such discourse, "the case has not yet been proven, at least to the satisfaction of the requisite authorities,"[25] this is doubly true in the case of genetically modified foods, which are often in fact offered as the environmental solution to the problems of pesticide poisoning that critics like Carson decried.[26] Thus, arguably even more than toxic discourse, its subset devoted to GMOs is "a discourse of allegation rather than of proof" that often employs a kind of "moral melodrama" to persuade its audience of the reality of the hazard.[27]

Buell's "toxic discourse" is clearly a species of Beck's "staging," that mediating role which cultural production has in making risks real, the analysis of which falls, of course, squarely in the arena of cultural critics' expertise. Against the presumption that "if we succeeded in turning everyone into an expert, risk conflicts would resolve themselves," Beck offers the staging of risk as a way to bring "complicating factors," like "different forms of non-knowing, contradictions among different experts and disciplines, ultimately the impossibility of making the unforeseeable foreseeable," to bear on the "clash of risk cultures" that characterizes world risk society.[28] As toxic discourse, *All Over Creation* figures staging quite explicitly, for Ozeki's imagined activists, the scruffy anarchist "Seeds of Resistance," carry out a variety of theatrical protests, from grocery store magic tricks, to roadside leafleting, to tofu-cream pie throwing, to digging up biotech crops in full biohazard gear. And the "moral melodrama" that is, according to Buell, characteristic of toxic discourse is very much in evidence. Indeed, one of the skits Ozeki's activists perform at the "Idaho Potato Party" (billed as an updated Boston Tea Party), is titled *The Tragedy of Cynaco the Evil Cyclops: A Morality Play in Three Acts*.[29] In a novel in which, as several critics and reviewers have pointed out, good and evil seem fairly cleanly to line up with the farmers and activists on the one hand and the biotech corporations on the other, this reference to a "morality play" feels suspiciously metafictional. But even if this David-and-Goliath narrative is fairly straightforward—putting Ozeki's novel firmly on the antibiotech side—her depiction of variously compatible and incompatible anti-GMO discourses, the characters' different responses to the question of what's wrong with genetically modified foods, makes the novel a dialogic staging of some of the common and often problematic arguments in these debates. As the characters respond, either explicitly or implicitly, to the (lack of) regulation of these products in North America, they point to the unknowns in "expert" and "lay" approaches to GM foods alike.[30]

Situated in the context of the regulatory analogizing of substantial equivalence, Ozeki's *All Over Creation* intervenes in its logic quite directly. The Seeds of Resistance, for example, perform a direct action in a grocery store in which they reveal to unsuspecting consumers that lurking in the seemingly innocent tomatoes and potatoes in their shopping carts are flounders and "bug poison." Reversing what they call the "perverted magic of biotechnology," they perform their own magic tricks, as when one of them removes a handkerchief from a potato only to reveal a large can of insecticide.[31] If readers are not taken in by the melodrama of these interactions—as when one of the protesters asks a young mother: "They're genetically engineering poisons into our potatoes these days. But they refuse to label it, so how are you supposed to know what you're feeding your baby?"[32]—we are likely at least left feeling uncomfortable about our own consumption. These theatrics thus have the pragmatic effect of eliciting the "yuck" response in the grocery store audience (and potentially also in the reader), but I also would argue that, in exposing and undoing substantial equivalence, these activists stage a mode of interpretation that can be applied to the novel more generally, as different characters attempt to represent potential nonequivalencies rendered invisible by this expert logic.

Rejecting the official regulatory "known" of the "analogous food," Ozeki's characters—and even the plot itself—seem to struggle with the question: to what might GMOs actually be "equivalent"? As Buell points out, toxic discourse is necessarily a product of its milieu, and thus it "may repress, fail to fulfill, or swerve away from itself according to the drag of other discourses with which it cross-pollinates."[33] The less certainty there is about the implications of the new technology, the more susceptible to "swerving" is the discourse. "Cross-pollination" is consequently rampant in *All Over Creation*, as the very "indeterminacy at the level of knowledge itself"[34] leads Ozeki's characters to fill the gap at the center of GM risk with what they *do* "know"—about morality and reproduction, about multiculturalism and diversity, about God and nature, about corporations and toxic chemicals—each of which produces a different version of the "unknown" of GMOs. In the process, *All Over Creation* highlights the dangers of ignoring the "gaps, discontinuities, and differences" in any analogical approach to GM foods.

INTRINSIC OBJECTIONS AND THEIR CROSS-POLLINATIONS

All Over Creation has caused some interpretive discomfort for readers, for although, like her previous novel, *My Year of Meats*, it is clearly "an ambitious, progressive novel that seeks to educate, persuade, anger, and motivate,"[35] her incorporation—and at times seeming endorsement—of what amount to fairly conservative responses to GMOs complicates what might otherwise seem a poster text for the liberal Left.[36] Thus, as reviewer Judith Beth Cohen notes, the novel evinces a "diagrammatic treatment of good and evil [that] is too neatly drawn," but also incorporates "zany characters [who] often compete with her political message," leaving Cohen wondering: "Does she really believe that we may be facing the end of the natural world, or is she simply having fun with the messianic self-righteousness of her activist creations?"[37] Disconcertingly, the answer seems to be both. Because the novel both stages and itself *is* toxic discourse, it is not always clear which positions are merely represented and which—if any—are actually endorsed. But certainly, the more "messianic" moments in the novel come when the characters express intrinsic moral objections to genetically modified foods, the type of objection that, as Robert Streiffer and Thomas Hedemann describe it, sees genetic engineering as "wrong because it involves the unnatural activity of moving genes across species boundaries," an objection generally seen either to be adequate in itself—"unnatural" being "wrong"—or indicative of "playing God."[38]

As the title, *All Over Creation*, suggests, this is an objection that Ozeki represents quite explicitly, and, in staging an awkward political coalition of anarchist activists and a Christian farmer, she offers both secular and religious versions—and suggests resonances between them. That such resonance is extant in anti-GMO discourse more generally is clear from the title of an essay that Ozeki reports inspired her novel, and that she has her fictional Cynaco representatives discover to their dismay: Michael Pollan's "Playing God in the Garden," published in the *New York Times Magazine* in 1998.[39] As Ozeki's corporate characters point out, "With its power to appeal to a broad-range demographic, that title was truly dangerous copy."[40] And this is true not only on the level of the potential for coalitional politics but also on the level of the troubling cross-pollination of theological and ecological concerns, both of which dangers *All Over Creation* stages explicitly. Ozeki's Seeds of Resistance, in the midst of a kind of pilgrimage across the United States in their biodiesel camper, turn toward Idaho when they encounter a newsletter, put out by former potato farmer turned heirloom seed saver and devout Christian Lloyd Fuller, that decries the production of genetically modified foods. Lloyd's theological objection to GMOs is clear: "Scientists do not understand Life, Itself, and when they meddle in its Creation, they trespass on God's domain. Beware the ungodly chimera they manufacture in their laboratories!"[41] And though some of the activists find Lloyd's Christian approach to be problematic ("All this God shit is way too heavy for me"),[42] their own ecological rhetoric contains elements of the theological as well, with "Nature" replacing "God": "Nature's own varieties are slowly dying out. Soon all we'll have are genetically modified mutants."[43] "Ungodly chimera" and "modified mutants" are the Frankenstein's monsters in these intrinsic objections, transforming "creation" in ominous, if unspecified, ways. For activists and fundamentalist alike, whatever these mutants might do (though this is also a concern, especially for the environmentalist Seeds), their mere manufacture is in itself objectionable.

The "unknown" of genetic engineering is here processed through the "known" of existing moral, ethical, and philosophical belief systems. And as the activists' invocation of "Nature's own varieties" suggests, intrinsic objections often posit a logic of "natural kinds" that are seen to be violated in genetic engineering. Describing similar responses in anti-GMO discourse more broadly, Donna Haraway notes: "Transgenic bordercrossing signifies serious challenges to the 'sanctity of life' for many members of Western cultures, which historically have been obsessed with racial purity, categories authorized by nature, and the well-defined self."[44] Haraway's list illustrates the potential for analogical sliding among these concerns—

from human to nonhuman, racial purity to plant purity—and, indeed, she finds, in the objections to the "mixing" of plant genes, "the unintended tones of fear of the alien and suspicion of the mixed," which, she argues, produces a "mystification of kind and purity akin to the doctrines of white racial hegemony" in the United States.[45] Though perhaps not wholly overt, such tones do enter into the discourse of Ozeki's "messianic" activists. Indeed, in a kind of textbook example of what might produce the discomfort that Haraway describes, one of the Seeds asserts: "There used to be this line that nature drew in her soil, which we simply weren't allowed to cross. A flounder, she said, cannot fuck a tomato."[46] Given the fact that neither flounders nor tomatoes reproduce in this manner, the confusion of human and nonhuman in "nature's" admonition seems clear.

Though such resonance indeed appears in the novel, however, it is in sharp contrast to what is otherwise a fairly straightforward embrace of multiculturalism and diversity, represented most obviously in the racially mixed character of the Fuller family. Lloyd, married to Japanese American Momoko, is himself responsible for some "mixing," and, as though aware that his focus on "purity" in plants might be misread as xenophobia, Lloyd further complicates the matter with yet another swerve, this time in support of "exotic" plants, which, unlike the genetically modified "mutants," are coded as "immigrants"—or, as he puts it "as immigrant as we are!"[47] As Ursula Heise has noted, Lloyd's support of exotics makes little ecological sense, for the incursion of "immigrant plants" into the Americas has been absolutely devastating for native plant life.[48] Also, while it is certainly true that large-scale industrial agriculture is characterized by monoculture, Lloyd, a farmer, surely does not want to celebrate the "diversity" of crabgrass and dandelions—unless this support for the exotic is understood as a desire to forestall charges of botanical xenophobia in his and the Seeds' fears of chimera and mutants.[49]

If these intrinsic objections to genetic modification seem to resonate uncomfortably with antimiscegenation discourse, the critic's discomfort is likely only to increase as these characters discuss what might befall the fruit of tomato and fish couplings. Here, Ozeki depicts controversy surrounding a technology that has become a potent symbol for anti-GMO discourse more generally, genetic use restriction technology (GURT), which activists (including those imagined in the novel) have dubbed the "Terminator." This technology, which was considered but never implemented by Monsanto, would make it impossible for genetically modified plants to produce viable seed, an innovation intended to guard surrounding plants from crossbreeding and to provide a kind of biological patent protection, as farmers would not be able to save or share seed from the crop. Though the issues related to the Terminator technology could be narrated in more "extrinsic" terms (of farmers' versus corporations' control over the means of production and reproduction, or of the ultimate commodification and commercialization of life), this is not precisely the discourse employed in *All Over Creation*. Rather, once again swerving from the intrinsic toward analogous moral arguments (and particularly the "sanctity of life" to which Haraway alludes), Lloyd describes the action of the technology: "This patent permits its owners to create a sterile seed by cleverly programming a plant's DNA to kill its own embryos . . . thereby, and in one ungodly stroke, breaking the sacred cycle of life itself."[50]

Describing seeds as embryos is not particular to Ozeki's character—Vandana Shiva has also described this technology as "programming the plant's DNA to kill its own embryos"[51]—but because Lloyd's "seed embryos" appear in a novel in which one of the major events around which the plot turns is his daughter Yumi's abortion, this discourse on "life itself" clearly resonates with "pro-life"—or antiabortion—discourse. Lloyd's anger regarding this event underlies much of the discussion of seeds in the novel, and the implied analogy becomes overt in a scene in which terminating a pregnancy becomes synonymous with terminating a plant's ability to produce viable seed. Here, Yumi, frustrated by the ways in which the activists are using her father to further their political goals, confronts them:

> I turned on Geek. "You've brainwashed my father! You're turning him into a goddamned poster boy for your politics—"
> "He's doing no such thing," Lloyd said. "This is not about politics. This is about life!"
> My face was burning. "Oh, for God's sake, Dad. It's just plants."
> Geek said, "Plants have a right to life, too."
> And then I lost it. . . . The two of them—the young radical environmentalist and the old fundamentalist farmer—made a ridiculous alliance, and I started to laugh. "Oh wow! That's the kind of pro-life bullshit that drove me out of here in the first place!"[52]

Yumi's assessment of this "ridiculous alliance"—"bullshit"—may be one the reader too is tempted to share—and to impute to Ozeki herself, particularly considering the otherwise progressive politics of the text. Geek later disavows this position, assuring Yumi that he is pro-choice; indeed, at another point, demonstrating the versatility of analogical thinking, he articulates genetic engineering in terms of choice: "We're trying to usurp the plant's choice."⁵³ But the plot itself also seems to suggest the doubling of "termination." As it turns out, the man who was responsible for Yumi's abortion at age fourteen, her then-history teacher Elliot Rhodes who had seduced her into statutory rape, is, years later, hired to do public relations for the Cynaco corporation. When Lloyd, suffering from a delirium related to his heart disease, calls Elliot "the Terminator," this confusion is represented as the delusion of a terminally ill man, but the reader too might note the uncanny coincidence in the same character occupying the roles of rapist and Cynaco spokesman.⁵⁴

OF "STEALTH SEEDS" AND THE LIMITS OF THE EXTRINSIC CRITIQUE

All Over Creation thus provides a catalog of critical discomforts, as seemingly progressive, and even, in the case of the Seeds, radical, objections to genetically modified food swerve into conservative social politics, suggesting that discursive cross-pollinations might produce "toxins" even more dangerous than the uncertain peril represented in the organisms themselves.⁵⁵ Cultural and literary critics, confronting these novel foods and uncomfortable with the biopolitical and theological implications of intrinsic arguments, might counter with something akin to that offered by Michael Hardt and Antonio Negri, who ultimately offer an extrinsic economic critique in place of an intrinsic moral one:

> Some have sounded the alarm that genetically modified Frankenfoods are endangering our health and disrupting the order of nature. They are opposed to experimenting with new plant varieties because they think that the authenticity of nature or the integrity of the seed must not be violated. To us this has the smell of a theological argument about purity. We maintain, in contrast, as we have argued at length already, that nature and life as a whole are always already artificial. . . . Like all monsters, genetically modified crops can be beneficial or harmful to society. . . . The primary issue, in other words, is not that humans are changing nature but that nature is ceasing to be common, that it is becoming private property and exclusively controlled by its new owners.⁵⁶

This is, of course, a version of the critique that Timothy Morton also mobilizes: "What's wrong with genetic engineering is that it turns life forms into private property to enrich huge corporations."⁵⁷ And, in making it, Hardt and Negri tap into several veins of critical commonplace, from the "theological argument about purity," to the notion that nature is "always already artificial," to the ambivalence of "monsters," which many of us have learned, from feminists and scholars of science and technology (Cixous's Medusa; Haraway's cyborg), to see as progressive figures of positive change, even of liberation or resistance. Indeed, Haraway has assured her readers that "transgenics are not the enemy."⁵⁸

There is certainly comfort in finding the "wrongness" of genetically modified foods, not in the organisms themselves, but in the context surrounding them. Not only is this context likely to be discursive—and therefore "readable" by literary and cultural critics—it is also easily placed in the realms of the cultural and the economic—arenas in which we have long exercised our expertise, if not always to the pleasure of anthropologists and economists.⁵⁹ Even if some of us might feel a nagging sense of foreboding in Hardt and Negri's "beneficial or harmful" (what if, after all, they are the latter?), as long as we remain in a world in which nature is "exclusively controlled by its new owners," an economic critique in and of itself accomplishes the biological and ecological critique ("endangering our health and disrupting the order of nature") without any necessary recourse to species boundaries or food purity or naturalness. That is, genetically modified food is still "wrong"; it is just wrong because corporate control is wrong, not because intervening in nature is wrong.

But, if "transgenics are not the enemy," if the enemy lies in the context rather than in the things themselves, what happens if and when the context changes? This is the discomforting question raised in recent work by political scientist Ronald J. Herring, who, chronicling his experiences in India, complicates the issues of politics, economics, and control. Herring, like Hardt and Negri, approaches GMOs extrinsically, focusing on the economic context in which the organisms operate rather than their moral rightness or wrongness. And like Hardt and Negri, Herring seems to oppose the privatization of life that corporate control of seeds might effect. But for Herring, control over these seeds is not so one-sided. Arguing that anti-GMO activists are elitist in their attempts to speak for farmers, Herring reports that, in India and in Brazil, genetically modified seeds have been successfully pirated by farmers themselves, and thus no longer belong to the "new [corporate] owners" to whom critics like Hardt and Negri object. Describing the "Robin Hoods" of seed propagation, Herring attempts to obviate the economic/property argument against GMOs. In India, Herring argues, farmers have "embraced the agrarian anarcho-capitalism of stealth seeds."⁶⁰

In the process of making his argument, Herring alludes to both issues raised by Ozeki's imagined activists—the Terminator and mutant mixings. Indeed, he cites *All Over Creation* in a footnote, in which he accuses Ozeki of perpetuating what he calls the "Terminator hoax," that is, the belief among some activists that genetic use restriction technology is presently in use.[61] References to the Terminator—in the novel and in activist discourse more generally—tend to bolster ecological objections to GMOs, since a technology that could render seeds sterile seems dangerous in and of itself, particularly if, as in some activist extrapolations, it were to escape the particular GM plants in which it has been engineered.[62] Debunking the "hoax" is thus a crucial first step in Herring's argument, and he follows it by using the anti-GMO discourse of purity as a foil against which to offer his own narrative. Where anti-GMO activists might describe the illegal influx of GM seeds from Argentina into Brazil as "contamination" and "pollution," Herring dubs these "illegal-immigrant seeds," implicitly endorsing this border-crossing "anarcho-capitalism" against the xenophobia of national regulation. And if anti-GMO activists cast GM seed purveyors as the Goliaths against which anti-GMO resisters are so many Davids, Herring turns these roles around: "In both Brazil and India, in different crops, technology developed by Monsanto was appropriated, redeployed and developed by small-scale entrepreneurs and farmers themselves, unmindful of TRIPS negotiators or NGO petitions."[63] When farmers actively pirate the technology and "illegal-immigrant seeds" sneak stealthily across borders, without regard for Monsanto's intellectual property, ought critics to join Herring in cheering on the "Robin Hoods" of biotech, stealing from the rich and giving the means of bountiful production to the poor?

Herring's argument is clearly controversial. Activists have speculated, for example, that "stealth seeds" are the result of a different set of "anarcho-capitalists," those very corporations that, when their products are denied access to markets, encourage piracy, thus contaminating seeds stocks so as to render any attempt to legislate against GMOs futile. In Marie-Monique Robin's documentary film *The World According to Monsanto* (2008), for example, she suggests that Paraguay's decision to legalize GM crops in 2005 was the result of just this kind of contamination, an argument that becomes especially persuasive when the government official she interviews seems to confirm it. As Robin concludes, "Whatever the origin, contraband has been profitable for Monsanto."[64] It can be difficult, in other words, to distinguish the Robin Hoods from the Monsanto wolf in sheep's clothing. I raise Herring's argument, then, not to suggest that the "democratization" of "stealth seeds" is a fait accompli but rather as a kind of thought experiment. Clearly, as in Lloyd's and the Seeds' arguments about a seed's "right to life" or "choice," the narrative of GMOs can be told in quite diverse registers, all of which swerve toward analogies with otherwise disparate human concerns—from miscegenation to immigration to abortion. But to focus purely on extrinsic economic concerns—to suggest, in effect, that control of genetically modified foods is analogous to other sorts of corporate control—is to leave aside the troubling question of whether there might still be something wrong with GMOs—whether such organisms are patented and policed by Monsanto or pirated and exchanged through "stealthy" networks of farmers—even if the existing discussion of intrinsic "wrongness" is dominated by a moral discourse that we might not wish to endorse. In this context, I find myself returning to Hardt and Negri's perhaps inadvertent gesture toward the uncertainties central to risk society: "Like all monsters, genetically modified crops can be beneficial or harmful." For, surely, even as we ought to take our critical responsibility seriously, performing discourse analysis to excavate the theological and the "pure," we must also be wary of becoming too complacent in this critique. Not all monsters will end up, like Cixous's Medusa, "not deadly" but "beautiful" and "laughing."[65] Even if they are—and even especially if they are—like the cyborg, "unfaithful" to their origins, some monsters may wreak havoc.[66]

That the theological or quasi-theological response to GMOs is prevalent is hardly surprising, for, as Slavoj Žižek has noted, in a context in which "science provides the security which was once guaranteed by religion," "in a curious inversion, religion is one of the possible places from which one can develop critical doubts about contemporary society (one of the 'sites of resistance,' as it were)."[67] Of course, the problem in risk society is precisely that neither science nor religion seems able to guarantee security. Expressing their fears of the unknown future hazards of genetically modified foods, Lloyd and the Seed activists map onto them the fears they know, and, as I have suggested, even the novel's plot seems itself to display a similar sort of symptom, as though fearing that readers might not accept the vileness of the corporation, Ozeki has tapped as well into our likely aversion to statutory rape. I am certainly in full agreement with Žižek and Hardt and Negri, all of whom argue that we cannot accept the "conservative (religious-humanist) ideology which all too often sustains this critique."[68] But, important as the economic critique that Hardt and Negri, Timothy Morton, and others have offered of genetic engineering is, it also risks replacing one set of "certainties" with another ("It's just not right," says Morton). What is missing both in theological intrinsic and in economic extrinsic arguments, then, is the incalculable, the risk that is reducible neither to conservative analogy nor to property rights—that is, in effect, not representable at all—those "gaps, discontinuities, and differences" inherent in analogy that mark the "unknown" that the "known" is marshaled to domesticate.

STAGING TOXIC LEGACIES AND UNKNOWN RISKS

Of course, Ozeki does not neglect the economic; indeed, this context is central to understanding the operations of risk and power in the novel. The title of the Seeds' morality play, *The Tragedy of Cynaco the Evil Cyclops*, is fairly descriptive of the anticorporate spirit of the novel as a whole, as Ozeki's characters are as concerned as Hardt and Negri are about the natural world "becoming private property," "exclusively controlled by its new owners." This anticorporate spirit is what leads to that "diagrammatic treatment of good and evil" to which Judith Beth Cohen objects. And, though several readers have pointed to the seeming authorial endorsement of Lloyd and the anti-GMO activists, these characters do not exhaust the "zany" cast of *All Over Creation*, nor need they represent the novel's final word on the wrongness of GM foods.[69] If the clearest indictment of agricultural biotechnology comes with more than a whiff of Hardt and Negri's "smell of a theological argument," a more ambivalent critique comes from a character who grapples directly with the risk/benefit analysis of marketability and toxicity, the conventional potato grower, Will. Farming potatoes on a large scale and with very tight margins, Will is more concerned with how to bring product to a willing market than he is with the finer points of God's or nature's designs. But as he and his wife, Cass, struggle with their own infertility, and as he watches those around him growing cancers as well as crops, Will provides an analysis of the calculable and incalculable risks associated with both conventional and biotech foods.

For Will, the appropriate analogical "known" for GM foods is indeed the analogous conventional food, but, as a potato farmer, he is acutely aware of the toxic quality of conventional food production.[70] As the industry and consumers in the novel demand the uniformity of the Burbank potato, a plant with "notoriously poor disease resistance" that "must be coddled with fertilizers and fungicides and other pharmaceuticals like an overbred poodle," a constant shower of "water and chemistry" must be applied to sustain agricultural production.[71] But these miracles of modern science necessary to maintain a monocultural ecosystem have unintended effects of which Will and other farmers are well aware, even if they cannot make precise, causal connections between their chemical exposures and subsequent health problems. As Will notes, "The wife's had a bout with cancer. . . . Mother-in-law died of it. Old Fuller down the road had a part of his colon removed. . . . Maybe its related. Maybe it ain't. And maybe if I was a scientist I could give you a better answer. But I'm just a farmer, so I can't say."[72]

As it turns out, though, Will is not "just a farmer"; he is also a Vietnam veteran, and as such he neatly embodies the toxic legacies associated with Cynaco/Monsanto. These legacies coupled with the relative power of its control over genetically modified seeds have made the Monsanto corporation something of a lightning rod for GMO sentiment, presented at once as the savior for the scarcities of the new millennium and the demon who might produce such scarcities, a polarization emblematized by the fact that, in 2013, Monsanto's chief technology officer, Robert Fraley, won the World Food Prize (an award variously compared to the Nobel Prize or the Oscars), and Monsanto was simultaneously voted the "Most Evil Corporation of the Year" by the readers of *Natural News*, beating out the Federal Reserve and BP by a very large margin.[73] Like Monsanto, Ozeki's "Cynaco" is a chemical-turned-life-sciences corporation that produced the defoliant Agent Orange in Vietnam, chemical herbicides like "Roundup®" (called "GroundUp" in the novel), as well as the NewLeaf™ (Ozeki's "NuLife") potatoes. As a result, activists, in life as in the novel, suggest that the ecological problems arising from this company's past chemical ventures should make us suspicious of the new products emerging from their laboratories. Quoting an activist publication, Ronald Herring derisively describes this critique: "Clubbed together with Dow Chemical, which together 'brought us Bhopal and Vietnam,' Monsanto [is] accused of planning to 'unleash genetic catastrophes.'"[74] Monsanto itself represents GMOs as a kind of redemption, a way to mend what chemicals have wrought. This narrative may be fairly tenuous in the case of "Roundup Ready®" plants—those engineered to survive otherwise toxic doses of the herbicide—but in the case of Bt plants—that is, plants, like the NewLeaf™ potato, that are engineered to contain the bacterial pesticide—the manufacturers promise fewer chemical inputs. It is not necessary to spray crops with toxins if the crops themselves are toxic—if only, so the argument goes, to insect life.[75]

Given the market for Burbank potatoes and the tight margins in farming, and given the known risks associated with chemicals, Will decides that he will accept the unknown risks of GMOs, at least as long as the market sustains it. In the process, he puts himself into the middle of the biotech controversy, policed by Cynaco, which sends a Pinkerton to monitor the farmers using its products, and targeted by the activists, who, in one of their theatrical protests, don biohazard gear to dig up some of his NuLife potatoes. But he is far from a GMO convert, and in the course of his own risk-benefit analysis, he highlights the industry's toxic past, and possible continuing effects:

> "In Vietnam, the government said spray and we sprayed. Never gave it another thought. Now I got this numbness in my arms that the doc says may be Agent Orange, only he can't tell for sure because of the exposure factor on the farm. It bugs me. Cynaco made Agent Orange for the army. They made GroundUp and now the Nu-Lifes, too."[76]

Will's response is thus not precisely that Cynaco is planning to "unleash genetic catastrophes," which suggests some calculation on the part of the corporation; rather, his fear is in the lack of calculation—perhaps itself a calculated tactic—that suggests the potential for unintended and often unacknowledged risks, past, present, and future. Will's objection is not precisely intrinsic, but nor is it purely extrinsic, for the uncertainty surrounding both process and product still leaves open the nagging possibility that there *is* something wrong with the organisms. He does not know whether Cynaco's Agent Orange and GroundUp are analogous to their NuLife potatoes, and this absence of knowledge makes him distinctly uncomfortable. Drawing a metaphor from the crop he farms, Ozeki describes the ominous feeling of uncertainty that accompanies his choice to plant these novel foods: "a doubt that had been eating at him, like a root rot, starting with the war and growing deeper with each disaster."[77]

Ultimately, in the novel as in life, the "root rot" that eats at Will also eats at the market more generally, and Will finds that he has problems finding buyers for his NuLife potatoes. In an example of successful boycott politics that Ulrich Beck might admire, as consumers in Europe and North America question what might be missing in the discourse of substantial equivalence, their "voting" dollars begin to "erod[e] the market, and prices were down. McCain, the largest Canadian potato producer, had decided to go GE-free and Frito-Lay had followed suit."[78] In the face of this market pressure, the NuLife potato (much like Monsanto's NewLeaf™, which was discontinued in response to similar consumer and processor complaints) is abandoned and the company's interest in the so-called Terminator technology disavowed. By the end of the novel, though Will has not converted his farm to organics, he does indicate that he plans to "check out" an opportunity to sell hay to a newly established organic dairy in the area.

In this response to the NuLife/NewLeaf™ potato, we might read something like a risk variation on "knowledge is power"—in this case, it is "nonknowledge" that is power. Doubt, suspicion, and discomfort triumph in the face of the cynical reassurances of a corporation that has, certainly historically, put profits ahead of human or environmental health and safety. Will's mobilization of analogy, and the one that seems, ultimately, to have swayed the market more generally, is thus to ask, in effect, are Bt potatoes and Roundup Ready canola analogous to potatoes and canola, or to PCBs and Agent Orange? Here, the suspicion is not just that Cynaco/Monsanto is using these products for ill-gotten gains, privatizing the food supply and producing a global dependency on its products, but that the products themselves might be dangerous.

But the shelving of a single biotech product does not represent the shelving of genetically modified foods generally. Shifting its product line, target market, and rhetoric, Cynaco (again like the corporation on which it is clearly modeled) decides to emphasize rather than downplay the differences between conventional and genetically modified foods. Duncan, Elliot Rhodes's boss, explains Cynaco's new tactic:

> "They want to take a step back and retool their entire presentation, targeting it to Asia and the Third World.... I've suggested 'Enlightened Compassion' as the motivating theme to drive the new campaign, which will focus exclusively on the human health benefits of GE crops, like Golden Rice and the other pharmaceutically enhanced lines."[79]

Because Monsanto did have a stake in Golden Rice and elected not to take out a patent (precisely to represent itself as enlightened and compassionate), for those for whom the problem with genetic modification is not the creation of novel life-forms but solely who has access to them, the Golden Rice phenomenon would seem to make at least these GMOs acceptable, despite the potential for risks not yet realized.[80] In the context of *All Over Creation*, however, "Enlightened Compassion" is corporate spin, not genuine humanitarianism.

MONSANTO'S "GLOBAL AGRICULTURAL FOOTPRINT"

All Over Creation largely leaves the story of Cynaco here. Elliot's position as PR representative is "terminated," and the narrative turns elsewhere, focusing ultimately on the domestic narrative of the Fuller family and the political future represented in the Seed activists' journey to the World Trade Organization protests held in Seattle in 2001. But Cynaco/Monsanto's advocacy of Golden Rice is part of a strategy we might trace beyond the novel, for it offers a more affirmative spin on the analogy that connects the corporation's chemical past to its biological present and future: As innovations in agricultural technology once produced a "green revolution" to respond to food shortages and famines, so too, the argument goes, will today's innovations produce a "gene" or "green-gene revolution" that will offer a technofix to ensure food security in our own risky times. As populations continue to grow and climate change makes for new agricultural uncertainties, Monsanto and other life sciences corporations are poised to augment the technologies of hybridization, monoculture, and chemical fertilizers, pesticides, and herbicides, all associated with increasing yields in the past, with genetically modified "climate ready" crops that can withstand the vagaries of our rapidly changing planet, thus saving untold numbers of people from starvation. The "gene revolution" is thus analogous to the "green revolution," a 2.0 technofix for a new global food crisis.

Though Monsanto's "enlightened compassion" branding can be seen in its more recent ad campaigns and slogans (currently "producing more; conserving more; improving lives"),[81] I would like to turn here to a different text that attempted to stage a dialogue on these issues, a 2008 "Zeitgeist" forum, now available on the Monsanto site as well as on YouTube, organized and moderated by Larry Brilliant of Google's philanthropic arm, Google.org, in which Google.org's Sonal Shah (formerly of Goldman Sachs), writer-activist-academic Michael Pollan, and Monsanto's CEO Hugh Grant were to debate the effect of climate change on the future of food.[82] Titled "Body 2.0: Creating a World That Can Feed Itself," the forum is clearly future-oriented, but Brilliant begins by framing the debate with two key figures from what might be conceived, now, as the "Body 1.0," that earlier moment of food scarcity, Paul Ehrlich, author of the *Population Bomb*, and Norm Borlaug, the so-called father of the green revolution. Positioned as the articulation of the problem and the realization of its solution respectively, these two ghostly figures preside over the events, their speculations on and adverting of disaster a past lesson on which we might, in our own time of projected crisis, draw. Google.org's "Zeitgeist" is presented merely as a locus for a timely conversation, and, in inviting Grant and Pollan, moderator Larry Brilliant brings together the iconic figures in contemporary debates about food in general and genetic engineering in particular. Pollan is on record about the uncertainties in Monsanto's GM products, for he describes buying, planting, and contemplating the risks of Monsanto's

NewLeaf potatoes, first in the *New York Times Magazine* article that Ozeki suggests inspired her own work, and then in his book, *The Botany of Desire* (2001), a tale that ends with his decision neither to consume them himself nor to feed them to unwitting potluck guests. The debate is thus notable for the polarizing figures involved, but I am also interested in the ways in which the venue and host of the event bring to light yet another analogy for genetically modified (and conventionally grown, for that matter) seeds: data or information.

Brilliant frames the debate by suggesting that its concerns are "serious sustainability" and "a power shift": "from whom and to whom this power has to shift to make critical decisions about the world's food." The "world that can feed itself" is, in the context of the forum, something of a synecdoche—in the sense of whole for the part—since "the world" ends up meaning something like "the developing world," those whose populations are threatening to overtake food supplies. In an image that evokes the yellow peril rhetoric of another era, Monsanto CEO Hugh Grant explains: "China is eating one America today—plus, minus. They're eating one America." And in a context in which climate change threatens yields, Monsanto promises to step into the breach, now armed with "climate ready" technologies that will, Grant promises, double yields while using a third of the resources: "That," says Grant, "is the cozy corner." The "Zeitgeist" forum is largely organized around such extrinsic questions as economy and environment, control and the "power shift," but the intrinsic question comes up, at least obliquely, in Brilliant's framing of the debate when he asks: "When the choices before us include complex issues, ethical, cultural, and scientific trade-offs—scientific uncertainties—who decides?" His aside on "uncertainties" may be buried, but it nonetheless draws attention to the gaps in knowledge that Grant gestures toward only dismissively: "The hard piece, I think, isn't science." Grant's phrasing is telling, for whether or not the actual science of genetically modified food is "hard," it is not "the hard piece" for Monsanto. Given the regulatory procedures of substantial equivalence, given the lack of labeling in the North American market, "who decides" is the Monsanto corporation.

Like Ozeki's Cynaco, Monsanto here presents itself as philanthropic, interested in technology transfers that will empower farmers worldwide, suggesting that the "power shift" for which Brilliant calls will be downward and democratizing. But, as Allison Carruth has suggested, "By century's end, food power," like political and economic power, "accrues to those who design, license, and oversee complex systems of information."[83] And Grant makes it clear that the power lies in the data. Drawing a telling analogy between Monsanto and Google, he notes: "We're generating half a trillion . . . data points a year. Which is kind of like this place. It's data generation, and then it's mining through that data to generate insights. . . . data mining on a gigantic scale. That leads to commercial success." The universal solvent—data—provides the means for amassing knowledge and power, thus capitalizing on yet another analogy for thinking about GMOs and "animals, plants, soils, and people" more generally, not as "living things" but as "plotted pieces of information."[84]

In many ways, these two corporations may seem necessarily at odds; if Monsanto is "the world's most evil corporation," Google's slogan is, famously, "Don't be evil." Google.org's place in the "Zeitgeist" debate is largely neutral and benignly philanthropic, providing a forum (with modest "liberal" intentions) without dictating a particular position in the debate—not unlike Google's own status as a search engine, a neutral tool for culling information, a device by which one might discern, from reasoned consideration of the divergent sites available, one's own position on a complex and contentious issue like GM foods. Indeed, Marie-Monique Robin's film *The World According to Monsanto*, a scathing critique of that corporation, stages Robin Googling as a central part of her role as investigative documentarian, one Goliath taking down the other as the activist Googles truth to power. But, at the "Zeitgeist" forum, this image of neutrality, represented in the moderator, the context, and the sponsor, falters a bit when Brilliant introduces Grant as his "friend," the two having met first at the Svalbard seed vault in Sweden, that seed bank of seed banks, sometimes described as the "world's agricultural hard drive," which proposes to defend global genetic resources against the ravages of war, climate, and time.[85]

If Google has become, as Josh McHugh puts it, "a company that, in effect, controls access to the Internet's natural resources," then it does offer something of an uncanny double to Monsanto, a company that certainly seems to aim to control access to the planet's agricultural resources.[86] And, as it turns out, the relationship of Google to Monsanto appears to be cozier than Grant's tentative analogy ("kind of like this place") might seem to suggest. Brilliant's and Grant's interest in the seed vault is, of course, of a piece with the philanthropic message of the "Zeitgeist" forum. A fail-safe for "doomsday" ("Doomsday Seed Vault Safeguards Our Food Supply," declared a 2012 headline in *National Geographic*), Svalbard is secure enough to withstand an asteroid impact or a nuclear bomb detonation. But, as with so many aspects of the slow apocalypse, the doomsday conceived is at once future and present. As Cary Fowler, executive director of Global Crop Diversity Trust, puts it, "Lots of people think that this vault is waiting for doomsday before we use it. But it's really a backup plan for seeds and crops. We are losing seed diversity every day and this is the insurance policy for that."[87] What it is that threatens the seeds depends largely on who is narrating the doomsday tale. Often cited are wars in developing countries where the infrastructure for local seed banks cannot withstand the pressure of violence. But, as environmental activists have also pointed out, arguably one of the major threats to seed variety is the consolidation and transformation of seeds (whether by hybridization or genetic engineering) by life sciences corporations like Monsanto. As the diversity of food plant varieties diminishes in an age of corporate megafarms, as genetically engineered plants risk contaminating conventional ones, often through basic natural processes of pollen dispersal, preserving seed diversity becomes a fail-safe that is appealing for both conservationists and Monsanto alike, the banked germ plasm offering a store of material that might mitigate the apocalyptic effects of total GMO contamination—and also be mined (or "data-mined") for future profit. Such would seem to be an argument for the financial interest of Svalbard for Monsanto, but though rumors of Monsanto's investment abound— and though another biotech firm, Sygenta, is a Svalbard donor—Hugh Grant's interest in the vault appears to be personal.

Similarly, though the Bill and Melinda Gates Foundation has provided extensive funding for Svalbard (and is a vocal proponent of genetically modified crops), Google's interest in the seed vault in particular and in agriculture more generally might seem tangential, but in the years since the "Zeitgeist" forum, Google's relationship to Monsanto has grown cozier. Again betting that "commercial success" comes in "the data," in 2013, Monsanto expanded what Grant calls its "global agricultural footprint" by purchasing the Climate Corporation, a company, originally founded by former Google employees and funded in part by Google Ventures, that collected previously free National Weather Service data and offered it for sale along with insurance products. This technology will now be a part of Monsanto's "Integrated Farming systems business"—a new addition to the "precision farming" system (with its proprietary patented GMO seeds and GPS software that allows "pinpoint accuracy in applying fertilizers and other chemical inputs") that Ozeki depicts her farmer, Will, struggling to understand in *All Over Creation*.[88] Monsanto's purchase of Climate Corporation dovetails with its proposed "climate ready" crops, those plants that would withstand the salinity of overworked soil or the extremes of temperature and precipitation, both attempts to insure against (and capitalize on) just that uninsurable volatility that Ulrich Beck suggests is characteristic of risk society. The Climate Corporation and the seed vault represent the means by which Monsanto hedges its bets on the future, adapting to the new "environments" of second modernity—the world of climate change and a potential "biotech Chernobyl."

Monsanto's attempts to insure against the weather, its seeming confidence in its ability to predict and engineer for the future, offer a telling contrast to the more precautionary spirit of Ruth Ozeki's novel. In the end, Yumi decides that the challenge is to "accept the responsibility and forgo the control," even if we are "powerless to forecast or control any of our outcomes."[89] Thus, though her activist Seeds also embrace the information analogy for seeds ("A pea's a program," Geek tells Frank),[90] they draw quite different implications from this metaphor. Whereas turning life into data leads Monsanto to mine and manipulate that data, the activist Seeds question both the ethics and the safety of the mastery or control implied: "Our syntax is still haphazard. Scrambled. It's a semiotic nightmare."[91] And *All Over Creation* provides, as well, an alternative model for archiving this data. The Svalbard seed vault is clearly a valuable hedge against doomsday, but it can also provide an alibi for complacency in the face of the slower apocalypse of corporate enclosures, biopollution, and monoculture. As the organization GRAIN suggests in its critique, "Faults in the Vault," "If governments were truly interested in conserving biodiversity for food and agriculture, they would . . . as a central priority, focus their efforts on supporting diversity in their countries' farms and markets rather than only betting on big centralised genebanks. This means leaving seeds in the hand of local farmers, with their active and innovative farming practices, respecting and promoting the rights of communities to conserve, produce, breed, exchange and sell seeds."[92] Such is closer to the model of information exchange developed by Ozeki's activists, who end the novel by proposing a small-scale and geographically diffuse "seed-library database" that "really takes advantage of the non-hierarchical networking potential of the Web," offering heirloom seeds for growers to "adopt"[93] and save for the future, a noncommercial form of propagation aiming to preserve the diverse materials previously contained in Momoko's small-scale "vault, full of treasures."[94]

SUBPOLITICS AND NARRATIVE

At one point in *All Over Creation*, prodigal daughter and adjunct English professor Yumi offers a tellingly literary spin on seeds as information or data: "Seeds tell the story of migrations and drifts, so if you learn to read them, they are very much like books."⁹⁵ The genome as the "book of life" is so familiar an analogy as to be almost a cliché, with genetic engineering either a slight revision (along the lines of hybridization) or the introduction of "monkeys on typewriters" producing a "semiotic nightmare" (as Geek suggests). For Yumi herself, this analogy is a way to bridge the gap between the two sides of her life, her agricultural past and her literary present. But Yumi also challenges this analogy by suggesting that there is "one big difference": "Book information is relevant only to human beings. It's expendable, really. As someone who has to teach for a living, I shouldn't be saying this, but the planet can do quite well without books. However, the information contained in a seed is a different story, entirely vital, pertaining to life itself."⁹⁶ Yumi is not Ozeki, but it is difficult not to read this passage as something of a metafictional tell, a moment in which the apprehension about the importance of books might not be limited to the character. Indeed, for the risk critic herself, faced with the materialities of GMOs, of pesticide poisoning and loss of crop diversity, it is tempting to see the preservation of actual heirloom, open-pollinated seeds, the work Ozeki's Geek performs when he sorts out Momoko's scattered collection, as the more vital and pressing ecological work, our own tropological wagers, our forays into the archives of risk, superfluous in the face of corporate seed grabs. The fact that this archival sorting is described in a novel, however, somewhat mitigates this prioritization of the seed, for the information in books—including in Ozeki's book—too is "vital, pertaining to life itself." How seeds are narrated, how understood—as commons or property, as embryos, immigrants, or possible Chernobyls—has a profound effect on how they are produced, disseminated, cultivated, and preserved.

Indeed, in "The Seeds of Our Stories," a talk delivered at Cal Poly in 2007, Ozeki offers a somewhat different analogy to emphasize the importance of "book information." Juxtaposing the work of scientists to the work of novelists, she suggests that the former are "very smart, very concerned, very impassioned and dedicated truth-seekers, often underfunded and cut off from reality, which is what happens when you are a research scientist working in a lab." "I don't fault them for this," she continues. "I am a novelist, and I spend most of my time underfunded and cut off from reality, too."⁹⁷ The problem, she suggests, is when science moves from research to implementation, from "pure" to "applied," and this is where there would seem to be, at least on the surface, a difference in science and fiction. Because, she notes, "there is no such thing as applied fiction," "there is a limit to the amount of damage I can do."⁹⁸ But this initial commonsense distinction does not hold. Ozeki recalls:

> I comforted myself with this thought. I repeated it often, "There's no such thing as applied fiction. There's no such thing as applied fiction." I repeated it during the 2000 presidential elections and the Florida recount, and the UN Secretary Council Hearings, when Colin Powell described Iraq's programs to manufacture weapons of mass destruction.
>
> I repeated it right up until the day that the United States started dropping bombs on Baghdad, and then I stopped. Because of course there is such a thing as applied fiction, and there's no limit to the amount of damage it can do.[99]

In Ozeki's analogical series, scientists are akin to novelists who are akin to politicians, and, if we close the circle, politicians end up akin to scientists. The question that connects GMOs to "terror"—what do we do in the face of risk?—ends up highlighting a key difference in the humility of precaution and the arrogance of preemption. The problem of GMOs may indeed, as Grant suggests, not be the science, but the regulatory practice that preemptively presumes similarity, ignoring the precautionary acknowledgment of difference.

For the reader interested in political intervention, the temptation to invert Ozeki's formulation is strong. If "there's no limit to the amount of damage it can do," then perhaps there is no limit to the amount of good a more salutary "applied fiction" might do. But *All Over Creation* resists the sort of didacticism that might facilitate its application. Though we might celebrate the Seeds' new database, we are left still with the awkward political implications of their moral stance. Farmer Will's historical analysis might be preferable, but his tentative solution lacks conviction. Indeed, *All Over Creation* does not leave us in the comfortable position of being able to identify an authorial mouthpiece. Staging the staging of GMO risk, Ozeki at once offers a representation of uncertainty and elicits a feeling of uncertainty, leaving her readers as uncomfortable as her characters are in the liminal spaces of nonknowledge. In "Flora Not Fauna," Susan McHugh offers a provocative metaphor for understanding *All Over Creation*, suggesting that Ozeki's use of a kind of roving point of view makes the novel "rhizomatic"—or, more concretely, "potato-like." Ozeki's "rhizomatic potato stories," she argues, offer a medium appropriate to what McHugh believes is the novel's message: "not to assert a single, true meaning but to represent a struggle over the many meanings for GMOs."[100] But if *All Over Creation* is a novelistic potato, it is clearly, as Will is, afflicted with the "root rot of doubt."

For Ulrich Beck, risk society is characterized by a kind of "division of expertise": "There are 'owners of the means of definition'—namely, scientists and judges—and citizens 'bereft of the means of definition,' who have the dependent status of 'laypersons.'"[101] But while literary and cultural critics are doubtless on the side of the laypersons, we are arguably not entirely "bereft of the means of definition." As in DeLillo's *White Noise* or Sinha's *Animal's People*, the potential arises, in Ozeki's *All Over Creation*, for an alternative form of grassroots political intervention akin to Beck's "subpolitics," or "the *decoupling of politics from government*," the activist staging of risk in the media that consequently shape and organize public opinion and public spending, leading to a kind of boycott democracy in which dollars are "votes."[102] This sort of consumerist response might be elicited by the anti-GMO discourses staged in Ozeki's *All Over Creation*. As a reviewer for the *Christian Science Monitor* suggested of the novel,

> Monsanto and other biochemical companies are concentrating hard on genetically modified food, while spraying herbicide on mandatory labeling laws to keep consumers worry-free. Hippies screaming about "Frankenspuds" are easy to weed, but a new literary threat may be harder for the industry to squash.
>
> Hog farmers are getting skinned alive by Annie Proulx's *That Old Ace in the Hole*. And now Ruth Ozeki takes a whack at genetic engineers with a wonderful new novel called *All Over Creation*. Along with Barbara Kingsolver, these politically oriented authors form a persuasive triumvirate. Their immense popularity among sophisticated women readers and book clubs means that the consumers who are most valuable to food manufacturers are being fed a diet high in anti-industry sentiments.[103]

Like the consumers that Ozeki's activists target in the grocery stores, the "sophisticated women" in book clubs promise to be the motor for a consumer boycott politics, and, if they are not swayed by "hippies screaming about Frankenspuds," they might find Ozeki's novelistic treatment of this terrain more palatable.

There are, as I suggested in the context of *Animal's People*, limitations in this approach to politics. In inspiring consumers to pressure McCain and Frito-Lay to go GE-free, the staging of GMO risk arguably simply pits one "Goliath" against another, without really grappling with the larger system out of which both "Goliaths" and their GMOs grow. Such an approach, which Beck, somewhat misleadingly in my view, calls "David *plus* Goliath,"[104] does not answer the problems that, for example, Hardt and Negri highlight in their critique of the economics of ownership and control. But whether or not Ozeki's novel itself goes far enough in imagining alternatives to the present biotech-industrial-agricultural complex, in staging this and other questions for her readers, Ozeki invites us to participate in the production of knowledge surrounding GM foods, a "responsibility" she ascribes to novelists and readers alike: "to question what is happening in this world, fashioned and controlled as it is by experts."[105] Or, as Beck argues, "when society has become a laboratory," "decisions concerning, and the monitoring of, technological progress become a collective problem."[106] Staging some of the uncomfortable cross-pollinations in anti-GMO discourse, *All Over Creation* offers opportunities to practice the discourse analysis in which we are expert, and excavating the unintended freight that anti-GMO discourse sometimes carries is a vital task. But the novel also challenges the critic to think through the limits of this kind of critique. Moralizing and totalizing as some "answers" to the wrongness of GM food might be, they are, as Will's more tentative example demonstrates, stopgaps for an absence of knowledge, a silence that critics also efface if we do not hear it beneath the seeming certainty of these arguments. In the face of unknowable risks, *All Over Creation* suggests that discomfort itself, unease, may be the critical stance necessary to confront the question "What's wrong with genetically modified foods?"

Analogy may be not only necessary but useful in conceptualizing genetically modified food; in any case, I am not sure there is an alternative to trying to understand the "unknown" except in terms of the "known." Commenting on the metaphor of "health" in thinking about ecosystems, Greg Garrard acknowledges that "metaphors and analogies are not only inevitable but necessary parts of environmental rhetoric"; given this, he suggests, they "should perhaps be assessed for their moral efficacy rather than their accuracy, coherence, or conceptual purity."[107] *All Over Creation* stages a variety of these metaphors and analogies, offering a way to begin to assess the moral freight, both efficacious and potentially detrimental, that such analogies bear. In so doing, the novel might remind us that this is hardly specific to the world of laypeople. Garrard goes on to cite Dana Phillips, who suggests that "any scientific hypothesis that conceals an analogy tends to devolve into a metaphor and wind up as a myth."[108] In the case of "substantial equivalence," analogy has not, to date, suffered this fate. But as scientists too have noted, the analogy of substantial equivalence contains, as Janet Bainbridge, writing for the Royal Society in 2001 noted, substantial "ambiguities": "The introduction of a transgene may, as a result of gene-gene and gene-environment interactions, cause unexpected collateral alterations in the phenotype of the novel organism, which the substantial equivalence approach might fail to disclose," or, "Hypothetically, an unintended hazard (toxicant, allergen or anti-nutrient) might not be detected."[109] Genes, like the "language" to which they are often compared, can be pleiotropic; when put into a larger system, they may produce phenotypic expressions that were not intended, predicted, or even predictable based on the component "parts" assembled. In other words, it is not only the language of "toxic discourse" that may swerve but also the "toxin" that such discourse is called upon to represent. As in Bannet's description of more literary approaches to analogy, "The gaps, the discontinuities, and the differences are as important as the likenessness."

Whether there will have been a "biotech Chernobyl" is an open question, answerable only in the speculative tense of the future anterior. Storing heirloom seeds in a nuclear-bomb-proof vault is one way to hedge bets on this potential biotech apocalypse, but this wager does not rise to the level of a robustly precautionary ethico-political practice. Such a practice would not merely ensure that a "backup" of life exists on an "agricultural hard drive," but also that the doomsday scenario that requires such a backup does not come to pass, a precautionary reading that requires humility in the face of uncertainties to come. In the end (as in the beginning), I am, of course, a literary critic, a layperson approaching the question of genetically modified foods, but given that these foods are, in North America, in the fields and on the grocery store shelves, unlabeled, their regulation based on that most literary of conceits, the analogy, I too feel the obligation that Kerry Whiteside suggests arises in precaution, the principle that "problematizes and opens to discussion the values that are implicit in the scientific framing of environmental issues," as well as in the framing of their discontents.[110] Reading *All Over Creation* in light of the larger debates on genetically modified foods provides at least one way to register critical discomforts, whether with the strategies of domesticating these organisms or with the uncertainties those strategies obscure.

FOUR | Letting Plastic Have Its Say; or, Plastic's Tell

The supreme court on Monday said the threat of plastic bags, which is choking lakes, ponds and urban sewer system, is bigger than the atom bomb for the next generation.
    —Dhananjay Mahapatra[1]

What Derrida once called "Western metaphysics" is now also a dust cloud of eroded top soil, a dying forest and what may now be the largest man-made feature detectable from space, the vast floating island of plastic debris that spans a large part of the Pacific ocean.
    —Timothy Clark[2]

Since this text here (private and public) does not come down to the content of its meaning, I abandon it more or less like an empty form, a mere container, one of those plastic packages that float (for how long?) on one of our beautiful rivers (why do I say "our"?). A miniscule simulacrum of nucleo-literary waste. . . .

"Things" don't "biodegrade" as one might wish or believe.
    —Jacques Derrida[3]

PLASTICITY, TOXICITY, AND TEMPORALITY

In 2008, the Los Angeles Department of Water and Power hit upon a solution to an unforeseen problem in the Ivanhoe Reservoir, which provides drinking water to the southern portion of the city. The problem was the presence in the water of the carcinogen bromate, formed when, in the open-to-the-sky pool, the bromide (which appears naturally) and chlorine (used in treatment) interact with sunlight. Cost-benefit analyses suggested that to cover the pool, with cloth or metal, would likely be too expensive and inefficient. And officials reassured residents that "dangers were minimal because bromate poses a small cancer risk only after consumed daily over a lifetime."[4] Nevertheless, the existence of a known carcinogen in the city's water supply was a cause for concern, and so a year after elevated levels were detected in the water, a remedy was found: To shield the water and prevent the volatilization of the bromide, engineers decided to float 6.5 million black plastic balls on top. The mood at the event, at least as the *Los Angeles Times* reported it, was ebullient: "Pebble-heavy 'plops' permeated the laughter of smiling onlookers. City Councilman Tom LaBonge shouted, 'For quality of water for all of Los Angeles!' with each of three balls he chucked into the water"; "'Water quality doesn't get more exciting than this,' Marina J. F. Busatto, a DWP [Department of Water and Power] biologist, said with a smile to a colleague as she helped slide ball-filled bags to the reservoir's edge." The only reported dissenter was a local resident, Marilyn Oliver, who offered a rather ominous "It looks like an oil spill," but "quickly added," "it's OK because it's temporary and the water quality is more important than the looks."

Of course, one cannot help but wonder whether those magic plastic balls, even while successfully shielding the water and preventing the production of one carcinogen, might themselves interact with water and sun. Surely anyone who has purchased a plastic bottle or baby toy in the last few years has had a crash course in plastics chemistry and endocrinology, as phthalates and bisphenol-A have moved out of the lab and into public parlance. Can we really believe that the new BPA-free plastics will be the end of this?[5] Aren't we justified in suspecting that these too might have some other hidden danger? (Plastic has turned out to be something like Chris Columbus's gremlins from the 1984 film, cute and lovable until you get it wet, let it stay up past midnight, expose it to sunlight, acidic liquids, or microwave ovens.) As though anticipating this response, the *Los Angeles Times* reports that the balls are "environmentally safe for drinking water and approved by NSF International, a government-sanctioned, nonprofit water quality organization." But in an age in which toxicity seems to change as one crosses borders (witness debates over the safety of BPA exposure, for instance), is "government-sanctioned" still a reassuring adjective?

Judging from the responses to the balls in the reservoir, the answer to this question is no. In a letter issued June 25, 2008, Marty Adams, director of water quality and operations for the Los Angeles Department of Water and Power, attempted to reassure those who expressed concerns about the safety of the plastic:

> Perhaps the greatest misinformation I have recently seen is the notion that the bird balls heat up and release toxic chemicals into the water.

This is just simply not true. Granted, there are many different types of plastics and recent news has focused on certain plastics leaching chemicals. . . . The balls are made of High Density Polyethylene—a long-time water industry product also used for pipelines, and a black version of the same clear product you buy your one gallon container of milk in (look on the bottom for the HDPE label). You will not find any legitimate news findings claiming that HDPE leaches chemicals into drinking water. . . . The balls are made to survive in a hot, sunlight environment without breaking down, and they are warranted for 10 years—twice the lifetime we are looking for.[6]

Of course, fear of contamination in and of itself can produce potent results, even in the absence of any revelation of actual harm. Adams's invocation of the wholesome and homey "container of milk" is doubtless targeted to dispel such fears. But Adams's concession that other plastics ("certain plastics") have been found to leach chemicals raises the possibility that high-density polyethylene (HDPE), that safe plastic, used so pervasively to contain such staples of life as water and milk, will not have been as inert as we presently believe it to be. Reassurances about the safety of HDPE are perhaps less comforting given what has happened to PET bottles in recent years. If, as a report from the Pacific Northwest Pollution Prevention Resource Center suggests, a "common plastic memory device" is "One, four, five and two, / all of these are good for you," recent research suggests that at least number 1 (PET) is not so "good for you" as it was once thought to be.[7] Though it contains no presently known endocrine disruptors, PET was found to leach "small amounts of one or more unknown compounds that mimic estrogen."[8] As Beck points out, in risk society, "What was judged 'safe' to swallow today, may be a 'cancer risk' in two years' time."[9]

But even if the bird balls in the Ivanhoe reservoir do not leach unsafe chemicals or pollutants—or at least, in the cost-benefit analysis that weighs such risks against the risks of bromate and the costs of a different covering material, not enough to warrant a more expensive solution—there is still the question of the stubborn and persistent presence of the balls themselves. When local resident Marilyn Oliver suggested that the first several thousand of 6.5 million plastic balls "looks like an oil spill," she was perhaps noting something more than aesthetic, given that most plastic has its origins in petroleum. And another onlooker, less sympathetic, might have pointed out the resemblance of the reservoir to that other, larger, reservoir, the Pacific Ocean, in which floating plastic presently makes up most of what is known as "the Pacific Garbage Patch," an area alarmingly described as the size, or sometimes (twice as alarmingly) twice the size, of the state of Texas.[10] This, we might hazard, could indeed be the ultimate locus for the Ivanhoe reservoir's black plastic balls, the use of which, as Oliver reassures herself, is "temporary." Though HDPE is one of the more recyclable plastics, and though Pankaj Parekh, director for water quality compliance for the DWP, in another attempt to placate citizen fears, informed a reporter for the *Times* that the balls would all be recycled,[11] it is not hard to imagine a few (hundred? thousand?) of these 6.5 million balls escaping that fate, either in ball form, or in an interim state like a plastic pellet, or in a new, recycled state, as plastic building material, toy, or flowerpot, likely the final stage in what is not as infinitely recyclable and "plastic" a substance as one might wish.

If risk criticism is necessarily concerned with the question of the archive—whether that is the archive of risk or the archive of hedges against risk, the critical piles or the doomsday seed vaults—integral to its own archival project is the chronicle of what was not meant to be archived, those unintended consequences of first modernity. In this project, perhaps no substance is as instructive as waste, excluded from the archive but, particularly in the case of plastic, building up as a material archive on the planet, and one that is, increasingly, haunting our cultural production as much as it is our ecological condition. When, exploring the possibilities of a cultural form of biodegradability, Derrida went looking for vehicles for its opposite, nonbiodegradability, it is no coincidence that he chose nuclear waste and plastics, both substances that seem, almost maliciously, to resist the test of time. Derrida imagines the future fate of his essay on the legacies of Paul de Man as akin to "one of those plastic packages," a "simulacrum of nucleo-literary waste," the eventual duration and consequences of which he can neither predict nor control.[12] Here, of course, plastic and radiation are the metaphorical vehicles for the tenor of the text, which is the question of the durability of the archive of Paul de Man's wartime journalism, not environmental concerns in their own right. But the metaphor could also be upcycled for more literal concerns. In "Responsibility, Biodegradability," for example, Tze-Yin Teo repurposes Derrida's "Biodegradables" for an ecological context by reading the figure of the biodegradable literally. Noting that, for Derrida, what is incalculable is the biodegradability of a document, Teo counters that, for literal wastes, "what cannot be mastered in the current environmental crisis is certainly not the eventual biodegradability of certain waste substances in time—the calculation for which is increasingly reliable as relevant technology progresses," but rather "the question of *an incalculable limitation of space.*"[13] Though Teo does not specify which of the "certain" waste substances she might mean, plastic would seem to fit nicely into this equation, as, refusing to biodegrade, it gradually usurps the space for others, human or nonhuman, who might be to come.

If plastic has, up to now, largely evaded what Derrida playfully calls "hermeneutic microorganisms," its material evasion of biological microorganisms has forced us to confront it, both as substance and text.[14] Plastic is, as Peter van Wyck has suggested of nuclear waste, not so much "matter out of place" (Mary Douglas's much-cited definition of pollution), but "matter *without* a place"[15]—both having duration that exceeds human imagination, and, as we are beginning to learn, forms of toxicity that are novel, difficult to contain or control. If the geologist of the Anthropocene might dig into the earth to discover the strata of this era, in between the layers that Timothy Morton has recently imagined—"1784, soot, 1945, Hiroshima, Nagasaki, plutonium"[16]—we might wedge "1907, creation of Bakelite," the "first fully synthetic polymer, made entirely of molecules that couldn't be found in nature."[17] Given that plastics do not biodegrade, all of the plastic produced is, conceivably, still with us—and the ever-growing, never-diminishing mass of it contributes to the sense of horror that plastic waste represents.

Unlike nuclear waste, which might appear in the form of contaminated clothing or laboratory equipment ("decks you can't stay on for more than a few minutes but which seem like other decks"), plastic is, of course, visible, but at the same time, surprisingly difficult to represent. Though "Texas" has been the metric in the popular press for representing the reach of the Garbage Patch, the question of whether it is one Texas or two is symptomatic of the fact that, as the National Oceanic and Atmospheric Administration (NOAA) points out on its website, there is no scientifically accepted way to determine its size or shape, given that it is made up of microplastic particles suspended in the water column.[18] In this sense the "Patch" is something like the "nebulous mass" that Don DeLillo's Jack Gladney is informed might be a result of his exposure to Nyodene D. When Jack asks the doctor whether it is called "nebulous" because "you can't get a clear picture of it," he is corrected: "We get very clear pictures. The imaging block takes the clearest pictures humanly possible. It's called a nebulous mass because it has no definite shape, form or limits."[19]

This problem of imaging—the sense that the Garbage Patch clearly exists, but, like the "mass" is "nebulous," with no "definite shape, form or limits"—is in fact true of the plastics problem more generally. As Jody Roberts suggests: "The spread of plastic has been more subtle, and it is perhaps for that reason that experts of all stripes missed it slipping into unintended places, travelling near and far such that nearly every cup of water from the ocean is likely to contain some plastic in some form of degradation and nearly every human subject found anywhere on the globe will likely bear the marks of a plastic modernity."[20] Indeed, Max Liboiron has suggested reviving the term "miasma" to describe the "additive and somewhat mysterious" health implications of plasticizers, those substances that render plastics more flexible or pliable.[21] In an irony that Ulrich Beck would surely appreciate, the "wonders" of first modernity—miracle plastics, with their additives and colorants, taking on infinite shapes, textures, and degrees of fire resistance—generate uncertain, mysterious, and illusive forces, comprehensible only by recourse to an earlier vocabulary and understanding of harm. And whatever sorts of plasticizing chemicals those banal objects may have had in the hand of the consumer, in the ocean, they gather more, becoming "sponges" for other pollutants like DDT, PCBs, fire retardants, BPA, yielding "tiny time bombs" for oceanic life.[22]

Plastic thus offers a challenge for representation with which artists, writers, and culture workers have recently been grappling, producing cultural artifacts that even scientific agencies have taken to employing in presenting plastic. Accompanying NOAA's "Science versus Myth" web page on the Pacific Garbage Patch is, not only a large rendering of the Pacific, with at least three "garbage patches" mapped, but also an image taken from graphic artist and photographer Chris Jordan's *Midway* series, which consists of photographs of dead, decaying birds, the very sign of the biodegradable, the stomachs of which are packed with the stubbornly nonbiodegradable plastic pieces they have picked up from the ocean and its environs. As Stacy Alaimo suggests of this series, it is the very juxtaposition of plastic's "eerily cheery"[23] color and the death that it has most certainly caused—and will go on causing, now that it is exposed for other wildlife to consume—that renders the scene so troubling: "Everyday, ostensibly benign, human stuff becomes nightmarish as it floats forever in the sea. The recognition that these banal objects, intended for momentary human use, pollute for eternity renders them surreally malevolent."[24] "Cheery" and "malevolent," the plastic seems to taunt us; Jordan's photographs capture, in an image, the juxtaposition of finitude, the very cycles of ecological, planetary life, and the virtually immortal plastic, products intended, in many cases, for a single "banal" use, the juxtaposition of muted, natural tones of decay and the grotesquely animated colors of persistence. And though "cheery" and "malevolent" are intended, perhaps, to be a description of the plastic's appearance, they also seem—and this may be the "surreal" part—to slide into describing the plastic's very nature, its essence, if not its personality or feelings.

Of course to say that plastic is "cheery" or "malevolent" is to anthropomorphize, to attribute human emotions to nonhuman things, and attention to cultural production surrounding plastic suggests that anthropomorphism and related figures of personification and prosopopoeia have risen to something of a tropological wager for those interested in representing plastic waste in a risk archive. Appearing in children's literature and Disney productions—Peter Rabbit, Mickey Mouse, Thomas the Tank Engine—anthropomorphism has been denigrated, at least traditionally, in literary and scientific circles alike, generally relegated to the category of mistake, not elevated to the category of trope. Paul de Man argues, for example, that anthropomorphism is "an identification on the level of substance" that "implies the constitution of specific entities prior to their confusion."[25] Here, the focus on "substance" is what makes this a "confusion" rather than a "figuration," for anthropomorphism seems to make an assertion about the essence of a thing, a process that, de Man argues, "freezes the infinite chain of tropological transformations," which, given the premium on such transformations in de Man's work, is clearly a problem.[26] In the sciences, of course, "substances" are hardly a problem, but anthropomorphism's "confusion" remains troubling, as it projects the emotional, intellectual, or interpersonal qualities of one substance ("the human") onto another, in the process failing to account for the actual substances under consideration.

But in recent work in what critics have taken to calling "new materialisms," this negative valance of anthropomorphism appears to be changing, particularly as those who might, formerly, have avoided substances turn to them. Jane Bennett, for example, has suggested that "we need to cultivate a bit of anthropomorphism—the idea that human agency has some echoes in nonhuman nature—to counter the narcissism of humans in charge of the world."[27] Reading nonhuman things as having human qualities is, for Bennett, less itself a matter of confusion than it is a symptom of it. She takes the example of a power grid: "To say that the grid's 'heart fluttered' or that it 'lives and dies by its own rules' is to anthropomorphize. But anthropomorphizing has . . . its virtues. Here it works to gesture toward the inadequacy of understanding the grid simply as a machine or a tool."[28] Given the unpredictability and uncontrollability of objects—a power grid or a plastic pellet—attributing human traits offers a way to grapple with the limits of understanding, suggesting an approach that we might, in the spirit of an earlier critical ambivalence, call a "*strategic* anthropomorphism," one that recognizes the pragmatic usefulness of anthropomorphism, even as it is cognizant of its limitations.[29]

In this chapter, I track the tropological wagers associated with animating plastic (whether in anthropomorphism or other, related figures—personification, prosopopoeia),[30] strategies of representation mobilized by activists, artists, and writers alike to capture and complicate the problem of plastics, navigating the intimate relationships that all life on the planet now has with the synthetic material that at once seems to mold itself into the shape of our dreams and desires and to persist as a nightmarish reminder of a past that refuses to remain past and a future that defies prediction. Anthropomorphizing plastic emphasizes the degree to which it *is* human, our creation and our responsibility, even as the strategy also complicates any simple demonizing of this substance that has become such a vital helpmeet to humankind. And if approaches to plastics generally seem to imply that human beings ought to address the problem—by banning bags in grocery stores or microbeads in facial scrubbers, recycling packaging or reservoir balls—these texts emphasize the enormity and ubiquity of a problem that we cannot not address by imagining the plastic as addressing us. Plastic thus becomes, not something to be turned to, but something that hails us, constituting us as subjects of risk society, just as much as we may be producers or consumers of plastic things.

PLASTIC MATTERS

In endorsing anthropomorphism, then, I join Jane Bennett and other advocates of the "new materialisms," critical practices that emphasize the agential capacity of nonhuman things, whether organic or synthetic, born or made. My additional modifier of "strategic," though, is intended to forestall a potential pitfall in some of this work. As amenable as I find the turns to "objects," "things," or "matter" in contemporary cultural and literary theory, as much as I view them to be a necessary and useful supplement to ecocriticism and ecocultural studies, I also find in them a tendency to level any distinction one might wish to make between different kinds of things, a tendency that can complicate an environmental ethics or politics that relies, often, on such delineation. This leveling is particularly symptomatic in the rhetorical technique of listing, which is common in this kind of work, whether in Timothy Morton's "wind harps, egg cups, cathedrals, underwater gas pipelines, poems, neutron stars, PDFs, and grains of salt,"[31] or Serenella Iovino's "stories, bodies, landscapes, bacteria, electric grids, quantum entanglements, waste dumps, animal testing, cyborgs, cheese, nuclear sites, oceanic plastic, art, time, nature,"[32] or Serpil Oppermann's "water, soil, stones, metals, minerals, bacteria, toxins, electricity, cells, molecules and atoms, and a vast array of nature's constituents as well as culture's trash and garbage."[33] Such interventions are directed at undermining what is perceived to be an artificial and problematic binary between nature and culture, a binary that critics perceive ecocriticism as endorsing. Patricia Yaeger has suggested, for instance, that ecocriticism is "so contaminated with nature as perfection or with a quest for organic truth that operating in its name is hard."[34] Given the direction of recent work in ecocriticism, I find Yaeger's fears about operating in its name to be a bit misplaced. Certainly those of us influenced by Donna Haraway, Bruno Latour, and Ulrich Beck are already working in the interstices between those problematic binary terms. But the effect of the rhetorical move of listing is often to blur that admittedly problematic nature/culture binary by implying sameness (all items in the lists have "agentic capacity," as Oppermann reminds us).

Of course, the fact that all matter is "lively" need not mean that all of its actions are the same—or that all matter and its actions should be valued in positive terms. It is precisely because all things have agency that one ought to distinguish among them. Cheese and nuclear sites, oceanic plastic and bacteria—these may be lively, but they differ in their life-supporting or biopolitical implications. And in making these sorts of distinctions among lively things, some recourse to—again, strategic—language of nature or the organic may be necessary. At the end of *Vibrant Matter*, Jane Bennett speculates on some of the ethico-political pitfalls of her own endorsement of a lively materialism, noting that, whatever its epistemological or theoretical problems, "the ideal of nature as the Wild continues to motivate some people to live more ecologically sustainable lives." And she notes that, even she herself, a convert to Latourian hybridity, is troubled by the normative principles that might (or might not) accompany the shift her book advocates: "One thing I have noticed is that as I shift from environmentalism to vital materialism, from a world of nature versus culture to a heterogeneous monism of vibrant bodies, I find the ground beneath my old ethical maxim, 'tread lightly on the earth,' to be less solid."[35] This is, in her estimation, a risk worth running, but it also might remind us that, in the case of plastic—that paradigmatically artificial substance—recourse to the natural or artificial has had tremendous pragmatic political effect.

Though one can and should certainly think plastic materiality differently—in terms less of nature/culture than of carcinogenesis or wildlife poisoning—a too hasty leveling of distinctions among different forms of "liveliness"—what promotes life and what denies it—risks ethico-political neutrality. When plastic is anthropomorphized by activists, it is generally, as in Alaimo's rendering, "malevolent"—even in its "cheeriness." As some critics have pointed out, though, a too categorical rejection of plastic denies the ubiquity and usefulness of this substance, a point Gay Hawkins makes in her analysis of "say no" campaigns against plastic grocery bags: "Catastrophic images of plastic bags as pollutants link them to the end of nature and fuel a sense of disgust and horror. There is no possibility that plastic bags might move us or enchant us."[36] As her phrasing suggests, Hawkins is interested in experimenting with the latter figuration. Thus, though "in a world represented as drowning in plastic bags, a concern with how plastic materiality is performed in various associations seems both indulgent and grotesque," she elects to let "plastic bags have their say."[37] Analyzing plastic bags, first as the target of bans, but then as useful in storing a wet bathing suit and aesthetically beautiful in the famous "plastic bag" sequence from the film *American Beauty* (1999), Hawkins animates different sides of the plastic bag's materiality and the multiple ways in which the bag interacts with human beings. In the process, Hawkins usefully reminds the environmentally conscious consumer of the ways in which we are complicit with plastic, much as we might wish to demonize this substance. Plastic bags are hardly exonerated in her essay, but they become, through use and aesthetic pleasure, something slightly more sympathetic and complex. Experimenting with a transformation in environmentalist affect, Hawkins asks, in her book-length investigation of waste, "Could it possibly be more 'environmentally friendly' to feel sympathy and ethical concern for rubbish rather than disgust and anxiety about its destructive impacts on nature?"[38] But what "environmentally friendly" would mean, in this context, is vague. Certainly, in her discussion of the plastic bag sequence in *American Beauty*, one gets the sense of the aesthetic possibilities of plastic waste, but what sort of environmental response ought to follow from that recognition is unclear.

To see how the twin moves of strategic anthropomorphism and strategic reinforcement of the nature/culture binary might work, I turn to another wind-tossed bag, the one tracked in Heal the Bay's "mockumentary" *The Majestic Plastic Bag* (2010).[39] This short film, released as part of the campaign to ban plastic bags in California, offers, not an anthropomorphism per se, but rather something of a beast-morphism, as the bag is treated as some sort of organic or natural object, a member of some elusive species. The film has the feel of a nature documentary, as the voiceover, in David Attenborough–style British accent provided by actor Jeremy Irons, narrates the bag's "migration" from its "release into the wild" on the "open plains of the asphalt jungle" to "its home, the Pacific Ocean." The bag is not pre-

3. From *The Majestic Plastic Bag*. (Courtesy of Heal the Bay. www.heal thebay.org.)

sented as malevolent—this is not the strangling plastic bag that maliciously kills ocean life (though later scenes suggest it may do this accidentally)—but as itself at risk: even in a city park, which "at first seems an idyllic place for the plastic bag," "danger lurks 'round every corner," as the bag is threatened by parks' employees and, "one of nature's most deadly killers," the teacup Yorkie.[40] Thus the film illustrates the ways in which an inanimate object like the plastic bag has material effects that might appear motivated, even if not the product of some sort of cognition. Plastic matter is, in other words, "lively," to borrow Bennett's adjective, even if it is not alive.

This playful confusion of nature and culture, animal and trash, is certainly emblematic of what Patricia Yaeger reminds us is a "world where molecular garbage has infiltrated earth, water, and air," in which "we cannot encounter the natural untouched or uncontaminated by human remains." But while Yaeger concludes, "Trash becomes nature, and nature becomes trash," the film's ironic tone suggests something else.[41] Its humor clearly derives from the beast-morphizing of the bag, but there is no actual confusion of substance, no sense that the bag is "really" migrating. Lest the viewer not perceive the dissonance in "naturalizing" plastic, the final frame informs us, "Plastic bags are not indigenous to the Pacific." In *The Majestic Plastic Bag*, then, the overt strategy—the one that supplies the humor—is in the confusion of nature and culture, pretending that the bag is on some sort of epic migration, but this strategy only works by reinforcing the opposite:

the bag is not natural, is not alive, is not organic; it is, in fact, very out of place in the ocean.

PLASTIC HAS ITS SAY

The dancing, migrating plastic bags in *American Beauty* and *The Majestic Plastic Bag* offer playful examples of the liveliness of matter, as well as the potential for this liveliness to be mobilized for political purpose. In both cases, though, this liveliness does not carry all the way to prosopopoeia. To create a context in which plastic bags *actually* "say" something requires a shift in genre, for, much as Gay Hawkins may want to animate the plastic bag, to give of sense of its multifaceted "thing-power" (a term she borrows from Bennett), she does not imagine the plastic bag "speaking," in the sense that a human being might speak. Plastics most certainly communicate in us and through us, transforming not only our physical landscape, our cultural mores, and our consumer expectations, but also the very material of our bodies. Whatever "thing-power" plastic objects might have, however, it is only in imaginative texts (fiction, poetry, drama, film) that a plastic bag might actually be allowed to "speak." Though the boy in *American Beauty* might read the floating, dancing plastic bag as "like a little kid begging me to play with it,"[42] the simile disavows the anthropomorphism. It is "like" a kid, but it is not a kid. It looks animate, but we know that the wind is just blowing it about such that it *looks* like it is dancing. It is not nature, not human; it is artificial, much like the family to which the boy who makes this small short within the film belongs. The bag is animate, but it demands little of the viewer beyond aesthetic observation.[43]

In the interest of pursuing the strategic possibilities of anthropomorphism further, here, I turn to a set of texts that take plastic animation past simile or metaphor, playfully giving voice to plastic, using the strategy of prosopopoeia, and thus literally "letting plastic have its say." These plastic things—a bag, appearing in Ramin Bahrani's short film *Plastic Bag* (2009); a garbage sea monster, imagined in Rachel Hope Allison's graphic novel *I'm Not a Plastic Bag* (2012); a ball, the quirky narrator of Karen Tei Yamashita's *Through the Arc of the Rain Forest* (1990); and a kind of diffuse miasma, in Adam Dickinson's poem "Hail" (2013)—all take on human qualities through speech, addressing the audience directly in order to render themselves objects of human concern. In each case, this is not, perhaps, true anthropomorphism—given the whimsy in each, there is no real confusion of "being" or "substance," no sense that plastic "really" could talk or think.

Rather, these texts animate plastic objects in order strategically to engage the viewer or reader in a different sort of ethical relationship to plastic. Presented not as cheerily malevolent but as oddly sympathetic and innocent, these plastic things observe the society of which they are products with a usefully defamiliarized eye that, in turn, requires a different sort of reflection on our own positions as subjects in risk society.

Such narratives are not new, of course. Indeed, the particular texts I consider here might be seen as contemporary iterations of a genre that came to prominence in the eighteenth century, the "it-narrative." As Jonathan Lamb describes it, "the narrative usually begins when a thing comes out of circulation and looks back at its career, sometimes nostalgically, but mostly with incomprehension, disapproval, or resentment."[44] Lamb's use of the word "thing" is important, for he distinguishes "things" from "objects"—the latter might be useful commodities that produce us as consumer-subjects; the former, however, act uncannily outside of the realm intelligibility, as those items we thought we threw away come back to remind us that, on a finite planet, "away" is itself an ideological rather than geographical concept.

I begin with another film in what I am tempted to describe as a growing subgenre of the plastic bag, Ramin Bahrani's *Plastic Bag* (2009). Like *The Majestic Plastic Bag*, *Plastic Bag* features celebrity voice-over. In this case, though, Werner Herzog's voice is not to represent some fictional ecocelebrity, but the bag itself. Known for his unconventional nature/culture documentaries (*Grizzly Man* [2005], *Encounters at the End of the World* [2007]), Herzog is an apt choice, for, as James Palmer remarks in his review of *Plastic Bag*, Herzog "lends a gravitas to this film that allows nihilistic observations and big questions to be asked by Plastic Bag without the slightest sense of pretention" since "we have heard this same voice asking similar questions in Herzog films."[45] Palmer's choice to turn the thing into properly named subject (not "the plastic bag" but "Plastic Bag") highlights the degree of anthropomorphism in the film, and alludes, perhaps, to the ways in which, as Paul de Man argues, anthropomorphism seems to "freeze the infinite chain of tropological transformations," yielding "no longer a proposition but a proper name, as when the metamorphosis in Ovid's stories culminates and halts in the singleness of a proper name, Narcissus or Daphne or whatever."[46] Though conceivably the real hedge on the bet of nonbiodegradability in this short, ephemeral film is the proper name "Werner Herzog," de Man's "or whatever" seems particularly appropriate to describing Plastic Bag, a character that hovers in the spaces between a "what" and a "who." That anthropomorphism produces the effect of a proper name is especially appropriate to the message of the film, for, as Derrida notes, a proper name militates against biodegradability ("the proper name—the proper name function—finally corresponds to this function of nonbiodegradability").[47] In the case of Plastic Bag, the purpose of the film is to extend plastic's cultural life to something closer to the nonbiodegradability of its material life, to make plastic waste appear in the cultural archive as it does in the ecological archive.

If *The Majestic Plastic Bag* seems to position its viewers as armchair nature-enthusiasts, observing our own detritus as uncannily possessed of an external, "natural" life of its own, it still allows us to remain passive observers. Not so in *Plastic Bag*, in which the Herzog-voiced protagonist addresses us directly, giving us insight into its interiority, its own existential crises arising from its expulsion from human society and its duration in the Pacific. Indeed, given that Herzog provides its voice, the bag is, strangely, gendered, and "his" relationship with the female consumer he calls his "maker" is at once oddly familial and romantic, producing a different model both for the relationship with plastic represented in the film and for the relationship of plastic objects to the viewer of the film. Plastic Bag comes to consciousness when he is opened by the checker at the grocery store, the opening of the bag a "breath" that imbues the plastic with life. This is a moment of misidentification, as the newly awakened Bag, in a comical enactment of commodity fetishism, mistakes the consumer for the creator, thus effacing, for himself and for us, the processes by which petroleum products like plastic are actually made. But the point seems also to target that casual consumer who, in choosing to accept the bag, produces it as waste, turning object into uncanny thing. In this context, it is perhaps not surprising that his "maker," the one who makes by consuming rather than producing, is female, reflecting an ideological feminization of consumption more generally. A good ecoconscious consumer, the maker reuses Plastic Bag a number of times, a process that climaxes when she fills him with ice for an injured ankle, an intimacy that Bag finds initially "shocking," but also alluring: "This brought me closer to her than ever before, my skin against her skin, my cold her warmth; I made her happy, and she made me happy. I thought we would be together forever." This almost Oedipal moment of transgression offers some comic relief, clearly, but it also might suggest to the consumer-viewer, presumably herself a user (and reuser) of plastic bags, a sense of her own unthinking reliance on—and potentially abuse of—plastic bags.

Describing the passage from "object" to "thing," Jonathan Lamb suggests that while the former is situated in the realm of the human, a part of the symbolic exchange of goods, the latter is external to this system, marked by its "irrelevance to any human system of value."[48] An object's transition to thing is, he argues, "not dialectical": "Once having made the change, things do not return as anthropomorphized items in the systems of exchange and symbolic labor."[49] It is this transformation that the film describes—and, in the process, undoes. Plastic Bag very quickly moves from useful commodity to uncanny thing, a process that the bag does not understand he has undergone. Even in the landfill, the bag imagines that "it must have been a mistake. . . . I imagined her crying, 'Where is he? Where is he?" And it is at this point that the bag comes to realize the different temporalities that govern plastic—the speed with which plastic turns from useful helpmeet to waste, and the uneven duration of its own lifespan against an organic environment: "The world decomposed. It was eaten by monsters, some too small for me to even see. Not me. I remained." Plastic Bag resembles, certainly, the monster in Mary Shelley's *Frankenstein*, a novel so often mobilized to imagine the havoc genetically modified food might wreak on its "makers," but unlike Frankenstein's monster, Plastic Bag never knowingly gives in to murderous wrath. The damage he might cause is collateral, as when, having found the Pacific Ocean, he is partially consumed by sea life: "Some ate pieces of me until they realized I was useless to them. I wonder where those little pieces are now." The ominously open-ended query reminds us that those pieces, wherever they are, persist, out of sight and mind, but still extant, now multiple miasmatic "things" that retain nonetheless the identity of the original.

Plastic's persistence becomes a source of great ambivalence in the film. Certainly, the nonbiodegradability of plastic is something that human beings tend to lament, even as we are, at the same time, grateful for a substance that is so resilient and resistant to mold, rot, rust, or decay. That the plastic itself might lament this—rather than, malevolently, glory in it—is something that we don't generally consider, given that plastic does not "lament" or "glory" about anything. In his review of *Plastic Bag*, James Palmer suggests that the film's protagonist is most analogous to those in the poems of Tennyson, "Ulysses" and "Tithonus," Plastic Bag's journey to what he calls the vortex taking on shades of the epic, his aging-but-not-dying a "cruel immortality."[50] The references here highlight the incongruity of the anthropomorphism, as epic heroes and princes become analogous to a flimsy piece of garbage, the waste of contemporary consumer culture. And Plastic Bag is not content to remain a thing. Though he finds temporary solace in the camaraderie of the vortex, a place where bags are "free and happy," the film's end leaves Plastic Bag still yearning to find his maker.

"And when I do," he tells us, "I will tell her one thing: I wish you had made me so that I could die."

Of course, Plastic Bag does not find his maker, but his message is passed along to viewers, proxies who have certainly themselves had similar relationships to plastic bags, and it is this form of address that is central to the film's impact. Describing Gaston Bachelard's argument about hybrid objects, Bill Brown notes that it "portrays the material object world as another medium through which we are constructed, by which subjects are, say, silently interpellated."[51] Plastic Bag, animated and ventriloquized, is, of course, not so "silent" in its "hailing" of the subject, but the Althusserian language is useful insofar as it highlights the ideological nature of the subject's constitution. The subject interpellated by Plastic Bag is individualized, consumerist, feminized, and as such a fairly familiar, even comfortable, fit for the viewer herself. The environmental subject produced is isolated, empowered perhaps in our individual acts of resistance (say, bringing reusable bags and therefore refusing to "make" another such bag), but the broader context of plastics—manufacturing, transport, recycling, disposal—is largely lost. Such individualized, consumer subject positions are hardly rare in environmental films—take for instance Al Gore's *An Inconvenient Truth* (2006), which capped off its devastating account of climate change with the modest suggestion that viewers swap their old incandescent lightbulbs for the new (mercury containing) compact fluorescents. The different scales on which the consumer and industry act produce something of the disjunction that Beck describes in the context of the Greenpeace protest of Shell: "'Those at the top' get the approval to dump in the Atlantic an oil rig filled with toxic waste, yet 'we down below' . . . have to save the world by dividing every teabag into three—paper, string, and leaves—and disposing of them separately."[52]

In this way, *Plastic Bag* may collude in what Beck has called the "individualization" of contemporary world risk society, a turn that is, for Beck's fairly liberal model of subpolitics, perhaps ambivalently positive (as individual consumers can "vote" with their dollars), but also quite limited. Hailing us, Plastic Bag calls for a response, but the sense of responsibility inculcated as a result remains locked in the claustrophobically contained dyad of consumer and her similarly individualized and anthropomorphized waste. Indeed, Plastic Bag goes further in explicitly resisting any form of collectivity. When he finally arrives in the "vortex," that plastic Garbage Patch in the Pacific, he discovers his "own kind" and with it the first-person plural: "I made it to the vortex. I was with my own kind. We covered the area of a small continent. We were free and happy." This shift to a collective "we" suggests a way to conceptualize a more widespread and pervasive plastic presence, but Plastic Bag is not content to join in a kind of hive mind. He discovers that "no one here thought about anything," the aggregate plastic "continent" seemingly incompatible with the vivification of prosopopoeial voice.

But individuality on the scale of the single plastic item is not the only way to animate plastic. A somewhat different strategy appears in Rachel Hope Allison's graphic novel *I'm Not a Plastic Bag*, in which a range of plastic and other waste items (a bag, a name tag, a tire, an umbrella) all make their way to the Pacific, there to agglomerate as a single organism that, serendipitously, ends up communicating via the text and icons on the surfaces of its pieces, a smiley face on the bag, "Hello" and "My Name Is" on the name tag.[53] Here, these objects-turned-things retain that residue of human voice which highlights their very incongruity. Though the words are in some ways accidental, over the course of the text, Allison presents the aggregate waste as a single organism, floating in the ocean like a prehistoric or extraterrestrial creature, luring birds to their doom—at once innocent and malicious in its cheery smiley face and friendly greetings. Allison's garbage creature is born, not in the process of consumption, but in those forgotten spaces of abjectification, thus emphasizing the "thingness" of its being. As the floating garbage assumes its shape—a tire forms one eye, an umbrella another; its plastic arms stretch down into the water—the text changes, again based on what is possible on discarded signs—"Hello"; "Come in We're Open"; "Welcome"; "Please"; "Nice Day"; "Thank You"; "Come Again"—in a seeming effort to communicate with the sea life the waste-creature encounters. By the end of the book, the particles of plastic having broken down—degraded, but not biodegraded—the same "face" appears in the sky above the water, smiling its eerie grin as a seagull flies below. Somewhere between *The Majestic Plastic Bag*'s beast-morphizing and *Plastic Bag*'s anthropomorphizing, Allison's strategy highlights the uncanny human/inhuman qualities of (plastic) waste. Commenting on her inspiration for the project, Allison recalls: "Like a lot of people, just learning that something so big and dangerous existed out in the ocean was kind of a shock, and the image of all that debris, trapped in such a remote part of the ocean, really stuck with me. That said, even some of the most disturbing pictures of the Garbage Patch also had a kind of beauty to them, and it got me thinking that the island itself might be a sympathetic character."[54] Like Plastic Bag in Bahrani's film, the creature here is innocent, even as the results of its presence are toxic, but in Allison's graphic novel, the problem is no longer a plastic bag per se. The "I" who is "not a plastic bag"

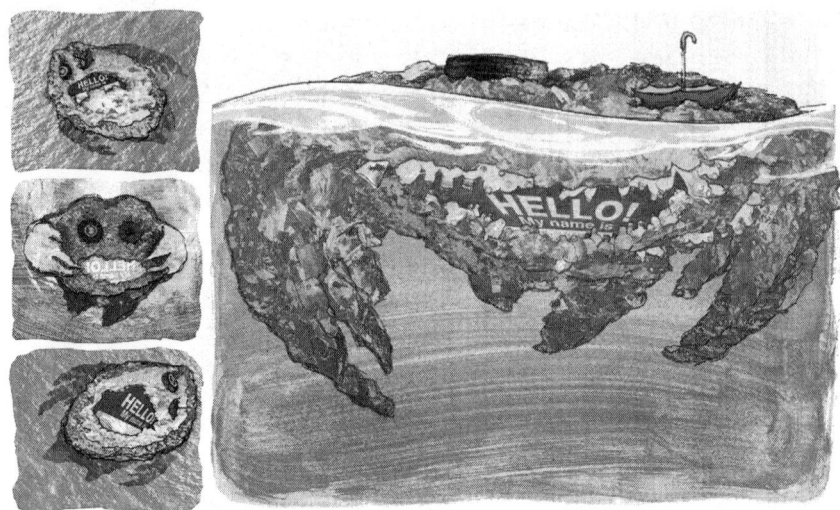

4. From *I'm Not a Plastic Bag.* (Copyright 2012 Rachel Allison. All Rights Reserved. Used with permission.)

is a larger waste problem, and one that haunts the scene even after the floating island has dispersed (but not disappeared).

"GLOBAL PLASTIC": IMAGINING THE COLLECTIVE

Bahrani's film and Allison's graphic novel offer tools for conceptualizing the plastic waste question—one can easily imagine, say, one of the Los Angeles reservoir's balls making a journey similar to that of Plastic Bag, though perhaps with a jauntier voice-over, or joining the aggregate waste creature in the Pacific, forming, say, an appealing dimple on its composite "face." These imaginings of the afterlives of plastic objects—their transits into inconvenient and uncanny "things"—help to forestall the cultural biodegradability that contrasts so dramatically with the ecological nonbiodegradability of plastic. Bringing these things back into circulation, *Plastic Bag* and *I'm Not a Plastic Bag* interpellate environmentalist-consumers, calling us to ethical responsibility—in the case of Allison's narrative, this consumer is figured quite explicitly in the paratexts, the forward by Jeff Corwin (executive producer and host of ABC's *Ocean Mysteries*) and afterword, "I'm Not an Ocean Polluter," which provide practical information on how to ameliorate the waste problem. Plastic objects are, in these texts, not "malevolent"; indeed, they are presented with the kind of "sympathy" and "ethical concern" that Gay Hawkins suggests might be more "environmentally friendly" than disgust or horror. In recounting the tragic fate of plastic—to be discarded with so little care or contemplation—these texts aim to develop our sense of responsibility to and for it. The anthropomorphism reminds us that there is a good deal of the human mixed up in plastic, emphasizing a relationship that we might wish to disavow, especially when the plastic moves from useful commodity to placeless synthetic matter. But these texts might highlight as well the representational challenges that remain. Rachel Hope Allison's composite creature begins to mitigate the individualism of the single plastic bag (and consumer), but arguably it still does not quite capture the elusive globality of plastic. Her garbage monster is modeled on the premise that the Garbage Patch is an "island," singular, representable, an image that avoids the troublingly nebulous qualities of its actual presentation in the ocean—a fact of which Jeff Corwin reminds us in the afterword when he describes the "patch" as a "trash stew." And graphic as such descriptions may be, they still do not take in the reach of plastic in all of its miasmatic ubiquity—qualities that make the global plastic phenomenon difficult to present as a "character."

Taking in plastic's global scale may seem, then, to require a strategy other than anthropomorphism, which tends, as I have suggested, to individualize, but I would like to turn, here, to a text that attempts to render these moves compatible, animating a single plastic "character," while still insisting on the elusively pervasive reach of plastic and other nonbiodegradable waste, Karen Tei Yamashita's magical realist novel *Through the Arc of the Rain Forest* (1990).[55] Set in Japan, Brazil, and the United States, this novel maps a series of transnational crossings, financial, cultural, and ecological, all of which culminate in an environmental cataclysm that is tragic for most of the characters involved. *Through the Rain Forest* is narrated by a mysteriously omnipotent figure whose identity is revealed (only very gradually over the course of the narrative) to be a ball made of a plastic material that appears to be the side effect of the burying of nonbiodegradable wastes around the world. This ball is in orbit near the forehead of a character named Kazumasa, a location that could limit its view, but it is also possessed, as it informs us, of a "clairvoyance" that allows it, in a first-person omniscient narrative, to map the movements of all of the diverse and far-flung characters, as a religious pilgrim in the north of Brazil makes a trek to the Amazon, or an American corporate CEO discovers the perfect product, or a jealous husband in São Paulo bemoans the business trips of his entrepreneurial wife.

The origins of this mysterious ball are certainly resonant with the first nuclear age, for, brought into being when a stray lightning bolt strikes the water near the home of Kazumasa (who is, at this point, a child in Japan), the ball appears as a kind of collateral piece of shrapnel, a molten portion of the earth's crust ejected in an unexpected and explosive moment. As Caroline Rody notes, "The thunder and 'flying mass of fire' that bring the ball to Kazumasa's head surely recall the atomic bombing of Japan and the mutations it bred."[56] But in this case the event in question is not the massive flash or fireball, not the subsequent rain of radiation, but "a tiny piece of flying debris," some minor bruising and swelling, and the addition of a personal satellite to Kazumasa's field of vision. This expulsion of debris is, as it turns out, revelatory of a very different set of ecological processes, which have developed unseen and unacknowledged (but no less materially) underground. As we learn as the story progresses, the ball is likely a piece of a much larger geological phenomenon, a kind of plastic layer in the earth's crust, formed, the scientists in the novel speculate, "for the most part within the last century, paralleling the development of the more common forms of plastic, polyurethane and Styrofoam. Enormous landfills of nonbiodegradable material buried under virtually every populated part of the Earth had undergone tremendous pressure, pushed ever farther into the lower layers of the Earth's mantle."[57] The ball, then, is a tip of this plastic iceberg, a side effect that signals, not the explosive effects of nuclear catastrophe, but the slower effects of an ecocatastrophe that has been happening, quietly and relentlessly, over the course of a century. Yamashita's novel thus offers us an image of what Myra Hird notes is basic to any form of waste disposal: "The management of waste ultimately fails. Fails to be contained, fails to be predictable, fails to be calculable, fails to be a technological problem (that can be eliminated), fails to be determinate."[58]

This new plastic layer of the earth's crust, the result of the geological period of the Anthropocene, eventually surfaces in a number of global locations, including the Amazonian rain forest in Brazil, a country to which Kazumasa and his ball immigrate in search of work and adventure. This geological formation, a giant sheet of plastic dubbed, by the locals, "the Matacão," is not only materially a hybrid (comprising nonbiodegradable material and the earth's natural forces), but it is hybrid as image as well, conjuring both the Garbage Patch, here brought by currents terrestrial rather than oceanic, and a kind of inversion of the "carbon sink" discourse surrounding Amazonia—in place of a carbon-absorbing forest, the earth's "lung," Yamashita figures a "garbage rise," a locus for the world's waste that, as Ursula Heise suggests, evokes "the simultaneous presence of natural ecosystems, cultures, and histories from across the world in one location."[59] Though the plastics problem, in our world as in the world of the novel, puts us all in the same boat (in the same plastic-filled ocean), Yamashita's Matacão might remind us that we are not all in the same proximity to hazard, not all as mobile or risk resilient; perhaps, to adapt Nixon's warning about life downwind, "We're all plasticized now, some sooner than others."

The ball's relationship to the Matacão, which Kazumasa imagines is its "true mother,"[60] is unclear, even to the ball itself. It is tempting, given the ball's description at one point as an "impudent planet,"[61] to read it as a metaphor for the earth, speaking to us with a voice that, as Heise notes, "emerges from the depths of geology," but it is a planet "strangely transformed," "half plastic and half rock, half waste and half raw material, and it orbits around a human head as if to signal the inevitability of anthropocentrism in even so fantastic a narrative strategy."[62] Indeed, in this case, the anthropomorphism seems to lend itself to the anthropocentrism, for the ball has far more insight into the lives and minds of the human characters (who, after all, seem to think as it does) than it does into the ecological actors, organic, synthetic, and somewhere in between—who are nonetheless quite lively—that play such a vital role in the events that ensue. The ball offers the charismatic anthropomorphized character with whom we might identify and sympathize, and the Matacão is the mysterious (even to the ball) substance that escapes such domestication. This relationship of plastic character to larger plastic waste phenomenon may initially seem akin to that figured in *Plastic Bag*. There, too, the "vortex" is an uncanny collective. But Yamashita's ball is not, as Plastic Bag is, an individual's waste, the persistence of which is to pique our individual consumer consciousness. Made of the same stuff as the Matacão, the ball, too, is representative of that larger agglomeration of waste, a phenomenon that comes from here and elsewhere and affects everyone, whether any individual forgoes a plastic bag at the grocery store or not. And, returning not as recognizable garbage but as uncanny sphere, the ball is a "thing" removed from the "object"—or objects—it might once have been, its lively operations incalculably unpredictable as a consequence.

Though Plastic Bag moves out to sea, seemingly positioned external to the system that produced it, the plastic waste in *Through the Arc of the Rain Forest* reappears in a new form, and, as it turns out, one available for returning to the system of commodity exchange. Having come, presumably, from the exploitation and transformation of previous natural resources, this new substance becomes a natural resource in its own right, subject to new mining and generating new revenue and industry, raw material for revolutionizing industry and global consumer trends alike. Though it is initially impermeable, eventually scientists manage, through a combination of acids and lasers, to mine and use the plastic. Once this technique is discovered, the Matacão is the ultimate "plastic" material, the perfect natural resource, submitting passively to the infinite whims of global consumer culture. Matacão plastic is so lifelike, so seemingly natural, that reproductions are often mistaken for the real thing: "The remarkable thing about Matacão plastic was its incredible ability to imitate anything. . . . Matacão plastic managed to recreate the natural glow, moisture, freshness—the very sensation of life."[63] This discovery inaugurates the "Plastics Age," as Matacão plastic is used in construction materials, in plastic foods (which are determined to be not only to be edible but virtually calorie-free), and in plastic surgeries—producing a "paradise of plastic delights" for modern global society.[64] But even while the novel tracks a giddy enthusiasm for this newly plastic planet, it also chronicles the side effects, for the Matacão seems to usher in further desertification of the Amazon, destroying habitat for wildlife and for locals who may formerly have made a living off of swidden agriculture or rubber tree tapping.

The exploitation of the Matacão provides an allegory for the (neo)colonial legacies of resource extraction in the global economy. For those for whom it is plastic, the Matacão is, at least initially, a tremendous windfall. As American CEO J. B. Tweep discovers, "An entire world could be created from it," and thus, in mining the material, his company, GGG, "had accomplished what no one before had been able to achieve: it had turned plastic into gold!"[65] But the miracle material is only lucrative for the alchemists at GGG. Though "the north of Brazil was a gold mine in plastic," "those who uncovered a piece of Matacão plastic in most cases did not have the technology to cut even a splinter of the stuff away from the mother lode."[66] And even the Brazilian government, which initially imposes legal and economic restrictions on research into Matacão plastics in an attempt to keep the profits from that research in the country, cannot capitalize on this natural resource within its borders. As congressional members argue, "Brazil had once before emptied its wealthy gold mines into the coffers of the Portuguese Crown and consequently financed the Industrial Revolution in England. This time, if there was any wealth to be had, it had better remain in Brazil."[67] This protectionism is, however, already legislated against by the global economy, for, as other congressional members scoff, "The treasure of the Matacão might, at best, make a small dent in their continuing interest payments to the International Monetary Fund."[68] Just as colonial powers used Brazilian resources to finance the metropolis, so the multinational corporation, GGG, uses the market for Matacão plastics to line the coffers of its New York office, the new metropolis for the neocolonial world. The Matacão, then, becomes the perfect material and stage for global capital. Not only is the global North able to dump its own waste on the global South, it is able to "recycle" this waste, and to make a handy profit doing so.

Accessing this plastic involves its own side effects, as the proprietary blend of acids and lasers produces a runoff that proves to be toxic. In a telling consequence, scientists find that rats exposed to the pollutant become analogous to the vampire capitalists extracting the material, as, in the fifth generation of those exposed, offspring "were found to develop fangs and tiny horns and an appetite for blood."[69] Further generations, though, are found to grow additional appendages, a mutation that characterizes the three-armed CEO J. B. Tweep himself, and thus the corporation, GGG, is able to spin these as potentially salutary, even therapeutic. Tweep's hopeful recasting of the runoff as salutary recalls Edward Teller's notorious queries regarding radiation. "Cesium 137 in the fallout, by affecting reproductive cells, will produce some mutations and abnormalities in future generations," Teller conceded. But, he continued: "This raises a question: are abnormalities harmful?"[70] Of course, those who are most likely to be exposed to the runoff—and thus the test cases in these experiments in mutation—are not likely to be those in charge of the industry itself. Indeed, attempting to downplay the dangers of its new pollutant, GGG assures the public that "disposal locations will not impact the social or environmental structure as they are usually spaces made vacant by the mining itself."[71] Even as the global plastic sheet, the result of the world's nonbiodegradable garbage, has rendered the soil fallow, killing forest and farm alike, its mining will leave the new by-product solidly contained in geographical place (at least for the time being). The context that the novel conjures is thus inescapably global, as the plastic not only literally brings the characters together but acts as a metaphorical vehicle for addressing global resource extraction of all kinds—and the social, economic, and environmental consequences that accompany it.

As it happens, the ball has a mysterious magnetic attraction to the Matacão that makes it instrumental in discovering deposits of the material, and thus both the ball and the novel's other characters are swept up in the larger narrative of capitalist development, at once innocent observers and complicit participants in the environmental and social devastation that ensue. This combination of innocence and complicity makes blaming any one actant—human or plastic—impossible. At one point, late in the novel, the ball reports that Kazumasa has "had enough of Matacão plastic": "He stared cross-eyed at me with a certain sinister irritation but eventually relented. It was not my fault. One did not simply separate oneself from a lifetime of proximity."[72] The plasticizing of the planet seems likely to continue, for, though Kazumasa believes "Matacão plastic was a finite resource," GGG, which has hired him and his ball to locate new deposits, has plans to send him outside of Brazil, "to Greenland, central Australia and Antarctica, not to mention every pocket of tropical forest within 20 degrees latitude of the equator."[73] Kazumasa's position here is very much like our own: inundated by plastic, we have certainly had enough, but, also like Kazumasa, we seem unable to avoid it, either as new products or as persistently nonbiodegradable waste.

But the novel also breaks this seeming stalemate, not by imagining some new environmental movement or technoscientific innovation, but via a rather magical solution produced by the process of evolution. Even as Kazumasa is looking at the ball with a "sinister irritation," the ball is looking back in a manner both "haggard" and "sad."[74] The cause of this helps to explain a cryptic mystery in the novel's frame, for the novel is not precisely narrated by the ball, but by a reincarnated memory of the ball. The ball itself, as it turns out, has become biodegradable. If the hybrid actants of garbage and geological forces produced Matacão plastic, such forces do not end with it, for a bacteria develops to fill the new ecological niche, and everything made of Matacão plastic—from facial remolds to food to buildings and clothes to the narrator itself—is consumed. The end of the novel closes the frame, the voice returning to the presumed present, in which the narrator informs us: "Now the memory is complete, and I bid you farewell. Whose memory you are asking? Whose indeed."[75] Though one might have presumed the memory to be the ball's, this final query suggests otherwise. As readers of the novel, having incorporated its narrative into ourselves, presumably the memory is, at minimum, ours, individuals who have, in joining in the process of reading, become a collective, potentially global in scope.

The magical solution that Yamashita imagines, whereby evolutionary transformations overcome plastic waste, would seem to be possible only in the space of a novel. In the real world, as we know, the addition and subtraction that Patricia Yaeger uses to describe the "techno-ocean"—it "subtracts sea creatures and adds trash"[76]—would seem to be relentlessly and mathematically predictable, making the total plasticizing of our planet a fairly safe bet. But as Bruno Latour reminds us: "If a maxim had to be stitched onto the flag of political ecology, it would not be, as some of its militants still believe, the lapidary formula 'Let us protect nature!' It would be a different one, much better suited to the continual surprises of its practice: 'No one knows what an environment can do . . .'"[77] Indeed, the media has been abuzz lately with a variety of techno- and eco-fix solutions to the problem of plastic waste that rival Yamashita's in magical implausibility. In 2011, *Nature* reported, in a story uncannily like that imagined in *Through the Arc of the Rain Forest*, that evolution itself has been acting in tandem with the plastic environment human beings have produced, for scientists have found, in the ocean, bacteria that seem to have adapted to eat the plastic, a fact that "might help explain why the amount of debris in the ocean has levelled off, despite continued pollution."[78] And, in 2012, the *Daily Mail* reported that scientists had found a fungus in the Ecuadorian rain forest that might "be used to break down plastic, and so rescue the world from one of its biggest man-made environmental threats."[79] Here, nature itself seems to offer a magical alchemy by which the problem of the nonbiodegradable can be overcome. Perhaps some future viewer will look on with condescending nostalgia at a film like *Plastic Bag* (of the sort one encounters in those born, say, after the end of the Cold War, who tend to look upon the first atomic age in this manner), plastic having achieved what Tennyson's Tithonus could not, having persuaded the gods to let him die—as if, to attach Derrida's phrase to a new referent, not to the "Paul de Man affair," but to the question of plastic waste, "Forget it, drop it, all of this is biodegradable."[80]

Truth may end up being stranger than fiction, it seems, but before we proclaim the triumph of evolutionary fixes and planetary resilience, we might take a further precautionary lesson from Yamashita. Matacão plastic is indeed eaten by bacteria, a magical ecological solution, but given our knowledge of side effects, we might linger on the penultimate paragraph of the novel. Here, the "old forest has returned once again," we are told, surely a reassuring moment for those who mourn its loss: "But," the final sentence reminds us, "it will never be the same again."[81] What residue remains of the Matacão? What, to borrow again from DeLillo's *White Noise*, will "happen to the toxic waste [or plastic] once it was eaten or to the microorganisms once they were finished eating"?[82] Invocations of similarly precautionary questions accompany Gwyneth Dickey Zaikab's article in *Nature*: "Specialist bacteria seem to be eating the plastic garbage we throw into the ocean," she reports, "but whether they're cleaning up our poisons or just passing them back up the food chain remains to be seen." According to the expert she cites, ecologist Mark Browne, this process of bacterial digestion could easily be "yet another mechanism for the particles of plastic that we throw away to potentially come back to haunt us."[83] Given Yamashita's use of the future anterior, the "memory" commissioned for us that, by the end of the novel has become the reader's, one must imagine that there is something (pre)cautionary in that final "But." Something remains; whether it is ecological or cultural, haunting our thoughts or our endocrine systems, embedded in our bodies or our brains—or both—remains to be seen.

FROM ANTHROPOMORPHISM TO PLASTI-MORPHISM?

Commenting on the inevitability of anthropomorphism, Timothy Morton offers a provocative inversion appropriate to a world of animate matter:

> I anthropomorphize. It's not that I anthropomorphize in some situations but not in others. . . . Just as I fail to avoid anthropomorphizing everything, so all entities whatsoever constantly translate other objects into their own terms. My back maps out a small backpomorphic slice of this tree that I'm leaning on. The strings of the wind harp stringpomorphize the wind. The wind windpomorphizes the temperature differentials between the mountains and the flat land. The mountains are shellpomorphic piles of chalk.[84]

These variations on the theme of "-morphizing" all give a sense of a world of lively actors. Thinking of plastic, we may inevitably anthropomorphize, a fact that the plastic characters I have discussed here take to comically hyperbolic extremes, but we do well to recall, as these texts also demonstrate, the extent to which plastic plasti-morphizes us in our own version of the plastics age. The existence of disposable plastic, products, as Gay Hawkins puts it, "made to be wasted,"[85] has produced the world in which we live in ways too pervasive to enumerate—food and beverage storage, computers, furniture, building materials, bird balls for reservoirs, the addition of plastic and subtraction of sea life. The biodegradability of that plastic is perhaps yet to come, but even if it were to break down, it may well be turned, not into the nourishing stuff of soil, but into the toxins that then reinvade our bodies, transforming us, as we are coming to see it has been doing for some time, biologically as well as socially or culturally. The pervasiveness of plastic, its "thing-power," requires that we ask: What happens when, in effect, we *are* a plastic bag, when plastic becomes part of our cellular structures, our endocrine systems, our cancer cells, our germ lines? What consequences might follow from recognizing what Stacy Alaimo has called the "trans-corporeality" of plastic waste?[86] What kind of subject is interpellated at once biologically and ideologically in such a moment?

I would like to close by turning to a poem that stages these queries, again by animating plastic, and by giving it a voice. The poem in question, titled (in a tantalizing allusion, perhaps, to Althusser) "Hail," is drawn from Adam Dickinson's extraordinary book *The Polymers* (2013), a text that plays with the chemical and ideological structures of a plastic world. "Hail" offers a series of sentence fragments, chopped into short lines that carry the movement of the poem forward through the landscapes and bodyscapes appropriate to its subject, as the banal disposable objects return as uncanny things, saluting the reader:

Hail

Hello from inside
the albatross
with a windproof lighter
and Japanese police tape.
Hello from staghorn
coral beds
waving at the beaked whale's
mistake,
all six square metres
of fertilizer bags.
Hello from can-opened
delta gators,
taxidermied
with twenty-five grocery sacks
and a Halloween Hulk mask.
Hello from the zipped-up
leatherback
who shat bits of rope for a month.
Hello from bacteria
making their germinal way
to the poles in the pockets
of packing foam.
Hello from low-density
polyethylene dropstones
glacially tilled

> by desiccated,
> bowel-obstructed camels.
> Hello from six-pack rings
> and chokeholds,
> from breast milk
> and cord blood,
> from microfibers
> rinsed through yoga pants
> and polyester fleece,
> biomagnifying predators
> strafing the treatment plants.
> Hello from acrylics
> in G.I. Joe.
> Hello from washed up
> fishnet thigh-highs
> and frog suits
> and egg cups
> and sperm.
> Hello.
>
> "Hail" from Adam Dickinson's *The Polymers*[87]

In "Hail," the speaker of the poem hails us with an anaphoric "Hello" coming from locations near and far, like a long-lost relative sending a postcard: "hello" from "inside the albatross," from "staghorn coral beds," from "can-opened / delta gators, / taxidermied / with twenty-five grocery sacks," from "the zipped-up leatherback," and ultimately from "bacteria," "breast milk," "cord blood," "sperm." And lest the environmentally minded consumer become too self-congratulatory in our purchases of clothing made from recycled milk jugs, say, the speaker reminds us that it and its fragments are also in "microfibers / rinsed through yoga pants / and polyester fleece"—in the air, the dust, the water. Moving from the ocean plastic—the albatross made iconic by Chris Jordan's *Midway* series, the beaked whale, which made international news when it was killed by ingesting plastic bags—back to the terrestrial consumer, his or her own body now housing something that is still capable of a "hail," these lines reverse that trajectory of Bahrani's plastic bag, back from the vortex and now pervasive. What is anthropomorphized is no longer a visible thing. It is pieces, a miasmatic "memory," conceived biologically and culturally, that speaks from everywhere and nowhere. It is hailing us like a long-lost friend, a traveler, one that we thought we had sent away, but that comes back, uninvited and unwanted. The final "Hello," the single word on the line, indicates that the hail may be from closer still.

Dickinson's poem offers a rendering of the real problem that NOAA highlights, that "manmade debris does not belong in our oceans and waterways." The NOAA website ends with the admonition that this problem "must be addressed," but this use of the passive voice may allow any of us individuals to abdicate our responsibility to address this problem. "Hail" turns that around—the plastic addresses us in a way that is uncannily familiar, as though it knows us, individually, as though we were somehow responsible for it. Our friendly helpmeet is still out there—inadvertently killing wildlife—as well as "in here"—altering our biochemistry, our ability to sustain ourselves, even quite materially in the sense of biological reproduction (breast milk, sperm), and it is calling to us, reminding us of our own complicity in its actions. We may very well, like Kazumasa, not be able to separate ourselves from "a lifetime of proximity," but a poem like "Hail" ensures that we also will not forget this fact, presumably in the hopes that we might mitigate future plastics production, consumption, and waste. In the face of the "nebulous mass" of the plastics problem, Dickinson's poem helps us to see something that is there, but that we normally either do not or cannot see, that miasmatic presence which plastic has on all scales, from the global to the intimately local.

PLASTIC'S TELL

Plastic is, of course, the preeminently morphable substance. Its very malleability, mutability, plasticity, renders it capable of representing a number of diverse forms, from plush toy animals to cars. As Bernadette Bensaude Vincent suggests, "Plastics are shapeless; they have pure potential for change and movement. They connote the magic of indefinite metamorphoses to such a degree that they lose their substance, their materiality, to become virtual reality."[88] Writing about this phenomenon in his short reflection on plastic, Roland Barthes notes: "At the sight of each terminal form (suitcase, brush, car-body, toy, fabric, tube, basin or paper), the mind does not cease from considering the original matter as an enigma. . . . And this amazement is a pleasurable one, since the scope of the transformations gives man the measure of his power, and since the very itinerary of plastic gives him the euphoria of a prestigious free-wheeling through Nature."[89] Barthes was writing in the 1950s, and while the amazement likely persists, the euphoria is decidedly tempered. Today, if we consider the original matter at all, in, say, our purchases of stuffed toys for cute birthday gifts, it is surely still an enigma, but the sense that plastic is "free-wheeling through Nature" is likely a decidedly less "pleasurable" one, the "terminal" forms all too terminal—persisting long after their use, in landfills or oceans.

In animating the plastic things that we might believe we have discarded, the texts I have examined here indeed perform a kind of anthropomorphism, but in retaining, in each case, the shape of the nonhuman thing (bag, ball, miasma), they also undo the seeming "magic" of plastic's morphing power. Plastics mimic all the way down, their endocrine-disrupting qualities coming precisely from their ability to take on the shape of hormones, misrecognized, in turn, by our own bodies. Plastic bags mimic jellyfish in the oceans, the favorite meal of leatherback turtles, or krill, crabs, or squid, which albatrosses feed to their chicks. Alarmingly, too, as Chelsea M. Rochman, Mark Anthony Browne, and their coauthors point out, our regulatory system tends to endorse mimicry: "In the United States, Europe, Australia, and Japan, plastics are classified as solid waste—so are treated in the same way as food scraps or grass clippings."[90] To classify plastic thusly is, however, not to let plastic truly have its say. As the authors go on to point out, in their own research they discovered that "at least 78% of priority pollutants listed by the EPA and 61% listed by the European Union are associated with plastic debris."[91] Were regulators listening to plastics, they might, as the authors argue, regulate them accordingly, thus producing a different future. Were plastics production and consumption to proceed as it has, "The planet will hold another 33 billion tonnes of plastic by 2050. This would fill 2.75 billion refuse-collection trucks, which would wrap around the planet roughly 800 times if placed end to end."[92] Reflecting what plastic is actually saying by regulating it as hazardous could, they speculate, reduce this to "just 4 billion tonnes"—still, clearly, a staggering amount of plastic.[93]

Certainly, in our own historical moment as in Barthes's, magical thinking about plastic persists, especially in the form of recycling, which consumers often see as a way to atone for plastic's sins, a sense captured well in Don DeLillo's account of a visit to the recycling center in *Underworld*: "Newsprint for newsprint, tin for tin, and we all feel better when we leave."[94] There are substances like this, not newsprint, but glass, for example, that can be recycled virtually infinitely. But the percentage of plastic waste that flows back into recycling is fairly low—the EPA reported it at 9% in 2012.[95] And even if it does make it to recycling, the most recyclable of plastics is not very "plastic" in this sense. In the case, say, of HDPE, the plastic used for the Los Angeles reservoir, the next form is likely the "terminal" one. But this does not, of course, mean that more will not be made. As it happened, Los Angeles' use of the balls was precedent-setting—they had not been used for drinking water reservoirs before (they were "bird balls" used at airports), but following the city's massive order, the manufacturer, Orange Products, a "pioneer in the plastic balls industry,"[96] quite understandably began marketing the balls for use in such facilities. As those of us on the ground continue carefully to separate our glass from our plastic, squinting to read the resin codes stamped on the bottom, Orange Products will be manufacturing these balls by the millions, warranted for ten years, likely extant for millennia. The "hook" sentence in Francisco Vara-Orta's article on the Ivanhoe reservoir—"The Los Angeles Department of Water and Power dropped the ball Monday. Actually, it dropped 400,000 of them"—ought perhaps to be taken more literally and seriously than Vara-Orta seems to have intended.[97]

In letting plastic have its say, the texts I have considered here offer a way to defamiliarize our own perspective, to look at plastic, not as something we can discard, but as something persistently present, insisting on our attention and care. Listening to plastic, in this case, means, not just observing its magic morphologies, the ways in which it can mimic "the very sensation of life,"[98] but also recognizing its differences, its persistence, its hazardous effects on life around it. In an age in which plastic is so pervasive, in which even the laudable project of the *My Plastic Free Life* blog must, by its very existence, rely on the plastics in a computer, we cannot simply demonize plastic.[99] But plastic is not just one thing among many. Plastic may well be lively, and it certainly acts—and in this way it is very like other "things"—but its participation in the world is something to be mitigated, treated differently, say, than the "water, soil, stones, metals, minerals, bacteria"[100] with which it nonetheless inevitably interacts. Plastic seems infinitely morphable—even anthropomorphable—but appearances can be deceiving. Plastic bags might seem to say, as in the title of Rachel Hope Allison's graphic novel, "I'm not a plastic bag"—I am a jellyfish, a dancer, an endangered species, a narrator—but an attentive reader must listen for what we might call plastic's "tell." Roland Barthes, amazed by the myriad and wondrous shapes that plastic can assume, but also wary of the way in which it cheapens, gives us a hint: "What best reveals [plastic] for what it is is the sound it gives, at once hollow and flat; its noise is its undoing."[101] Letting plastic have its say, the cultural texts I consider here offer another way to imagine the "sound it gives," for the stories that these plastic objects tell are also a "noise" that might help lead to plastic's undoing.

FIVE | The Port Radium Paradigm;
or, Fukushima in a Changing Climate

The spooky thing is, we discover global warming precisely when it's already here. It is like realizing that for some time you had been conducting your business in the expanding sphere of a slow-motion nuclear bomb.
—Timothy Morton[1]

It's macabre to compare two kinds of risks: climate change and atomic energy.
—Ulrich Beck[2]

PARADIGMS OLD AND NEW

When, in her environmental memoir *Refuge* (1991), Terry Tempest Williams braids the narrative of the rising waters of the Great Salt Lake with the narrative of her mother's struggle with cancer, she acknowledges that though both are "environmental," only the latter has a human cause. As her final chapter, "The Clan of One-Breasted Women," suggests, the cancers in her family are at least potentially linked to their status as downwinders in the era of above-ground atomic testing. In the case of the lake, however, the culprit is rain: "There is no one to blame, nothing to fight. No developer with a dream of condominiums. No toxic waste dump that would threaten the birds. Not even a single dam on the Bear River to oppose. Only a simple natural phenomenon: the rise of Great Salt Lake."[3] In the context of the book, of course, this implied binary—between disasters anthropogenic and not—is somewhat undercut. The level of the lake and the status of the refuge are themselves artifacts of past and present practices of land management; nevertheless, the basic fact of precipitation appears to be on the order of brute nature. Even as *Refuge* came to print, however, this sort of presumption was in the process of changing, such that the questions Williams raises about the causes of her mother's and grandmothers' cancers—"I cannot prove that [they] . . . developed cancer from nuclear fallout in Utah. But

I can't prove they didn't"[4]—were being raised about the very atmosphere from which otherwordly rains seem to fall, as Bill McKibben, in his seminal book *The End of Nature* (1989), was alerting us to the fact that "the temperature and rainfall are no longer to be entirely the work of some separate, uncivilizable force, but instead in part a product of our habits, our economies, our ways of life."[5] Here, climate and nuclear shift, becoming, potentially, analogous rather than opposed, both the product of a first modernity that is generating the new environment of the second, a relationship captured in Timothy Morton's provocative simile: "It is like realizing that for some time you had been conducting your business in the expanding sphere of a slow-motion nuclear bomb."[6]

For Morton, the simile is important—global warming is *like* the bomb—for the bomb ends up being central to his notion of the "end of the world" that we presently inhabit. But if, here, climate change resembles the bomb—insofar as it is catastrophically destructive—the simile now jarring us out of the complacency enabled by its slow motion, others have suggested it might be like the bomb in taking precedence over all other threats, including the nuclear that Morton deploys as vehicle. In this new climate, nuclear reemerges, not as the existential threat to end life as we know it, but as the savior that will lead us out of these dark days—literally, as we heard in my adopted home province of Ontario, "keeping the lights on"—its image refurbished, now as a "clean and green" alternative to fossil fuels, precisely because of its relatively low contribution of greenhouse gases to the atmosphere.[7] In his more recent account of climate change, *Eaarth* (2010), McKibben notes that, whenever he gives speeches about global warming, he can count on the fact that someone will lead off the question period by "approach[ing] the microphone and ask[ing] with barely concealed glee if building more reactors isn't the 'solution' to the problem."[8] McKibben imagines that the questioner's thought is that McKibben is "an environmentalist, and hence I must oppose nuclear power, and hence aren't I a moron,"[9] and McKibben goes on to point to the inefficiencies, dangers, and expenses of nuclear power to show that he is not. But in fact the nuclear industry has become so adept at recasting itself that being an environmentalist and advocating nuclear power are no longer ipso facto opposed. A number of those activists and thinkers formerly associated with environmentalism—from Greenpeace cofounder Patrick Moore, to Stewart Brand of *Whole Earth Catalog* (in regular publication from 1968 to 1972), to James Lovelock of the Gaia hypothesis—have come out in favor of nuclear power as an expedient antidote to carbon-driven climate change.[10] Indeed Patrick Moore has recently appeared in an advertising campaign for the

Nuclear Energy Institute, an agency the mandate of which is to "foster the beneficial uses of nuclear technology before Congress, the White House and executive branch agencies, federal regulators, and state policy forums," which asks, presumably rhetorically: "How can you be an environmentalist and *not* support nuclear energy?"[11]

Such are the strange twists in the narratives of world risk society—narratives that seem always to end too soon. As in the folksy wisdom when it comes to the weather, "If you don't like the climate on nuclear energy, wait a while," for the nuclear question seems perpetually open, as catastrophic releases of radiation fade from public memory, if not from physical bodies or environments. Of course, the reason that Patrick Moore could, just a few years after the Fukushima accident—the area still thoroughly contaminated; the melted reactors so toxic that robots are only now being developed and deployed to assess the damage—argue that nuclear energy is the solution to climate change is that climate change is very much a problem, and one on a scale that seems to trump all other concerns. At 400 parts per million of CO_2, we have now significantly surpassed Bill McKibben's "350" mark, the level at which his current 350.org organization aims. A 2014 National Climate Assessment report confirmed the sense that many of us already might have had, having weathered hot summers, icy winters, and megastorms, that "climate change, once considered an issue for a distant future, has moved firmly into the present."[12]

There is no doubt that climate change is a paradigm case in world risk society, a runaway unintended side effect of fossil fuel consumption that continues unabated. Though a number of unspecified "innovations in the life sciences" also informed the *Bulletin of the Atomic Scientists*' recasting of the Doomsday Clock in the second nuclear age, climate was the *Bulletin*'s chief concern. It is only logical, then, that *Risk Criticism* would conclude with a turn to climate change. In the spirit of the Clock, though, I will not be leaving the nuclear behind—for though the discourse surrounding the nuclear has changed, the contemporary post–Cold War world has not left the nuclear behind either, and the reflexive qualities of risk society and its attendant criticism are, in this context, vital. The *Bulletin*'s decision not to call the contemporary era the climate age or the Anthropocene may simply be the result of that organ's fidelity to its origins; it may also be a refusal of the progress narrative that says once we had one set of risks and now we have another. When Naomi Klein proclaims, in her recent book, that climate change "changes everything," this is as much aspirational as it is descriptive; that is, she hopes it changes everything for the better, as we suddenly refuse the capitalist global energy glut in which we presently are complicit. In the spirit of the change for which she hopes, we ought to be mindful of the ways in which climate change necessarily effects a reassessment, but we should also be cognizant of the ways in which what seems new can be the old dressed up in change's clothing (as in the carbon markets that Klein takes to task in her book).

Climate change is an environmental crisis of global proportions, and one that clearly necessitates the kind of cosmopolitan thinking that Beck insists may follow from being at risk. But the very indisputability of climate concerns—though there are some persistent deniers, they tend to be out of the scientific mainstream—makes climate susceptible to two related problems. The first, what I would call "carbon fixation," is not the salutary "fixing" of carbon again in biomass, but a cultural fixation that renders all else—toxics, waste, GMOs, nuclear—secondary in the face of this single existential threat. This is something along the lines of what Erik Swyngedouw has dubbed CO_2 "fetishism," which he associates with a "post-political" discourse of sustainability that ends up sustaining neoliberal business as usual.[13] The problem is not, of course, attending to climate or to CO_2; reducing greenhouse gases is absolutely necessary to a livable future. There is a danger, though, in making climate the new synecdoche for risk in our times: thinking this risk in the absence of others (or in priority above all others) makes it susceptible to what Sikina Jinnah calls "climate bandwagoning," a "type of strategic linkage that involves the purposeful expansion of regime mission to include new climate-oriented goals that linking agents believe will further their own agendas, regardless of whether such linkages detract from the common good."[14] The nuclear industry's self-presentation as carbon-neutral seems a textbook example of this latter move, as, to borrow again from Swyngedouw, "The nuclear 'fix' is now increasingly staged (and will undoubtedly be implemented) as one of the possible remedies to save both climate and capital."[15]

I am cognizant of the risk I am likely taking here in criticizing the use (and abuse) of climate discourse. In the highly fraught and deeply consequential debates on climate change, nuance can easily be lost. I would like to take a moment for a precautionary disclaimer: I am no climate change denier, nor do I mean to downplay its very real and devastating global impact. Quite the contrary. I am concerned, however, about the ways in which our much-warranted focus on this issue takes our environmentalist eyes off of the other risks that persist, perhaps desensitizing us to the perils of genetic engineering (with climate-ready crops) or nuclear power, now seen as those tough choices we have to make to mitigate what really matters.

What sort of challenge Fukushima will have provided to the climate-friendly staging of the nuclear is not yet clear, but certainly nuclear proponents have spun it as a mere speed bump on the road to increased nuclear energy often called the "nuclear renaissance." The Nuclear Energy Institute's slogan, "Nuclear is clean air energy," demonstrates the power of climate bandwagoning synecdoche, as the premium on CO_2, now recast as the sole air pollutant, occludes the fact that nuclear may not be "clean water" or "clean soil" energy, particularly for those at the mining and disposal ends of the nuclear fuel cycle, but also, arguably, all along the route of the "peaceful" atom, a euphemism for the slow violence that nuclear has long represented.

Climate change may indeed change everything, but this is precisely why its resignifying power must be tracked carefully, as its very novelty and scale seem to induce—or at least collude in—the collective amnesia regarding the hazards that, just a few decades ago, seemed to promise the extinction of life on earth. As Tom Cohen has argued, with reference back to Derrida's nuclear criticism, "Climate change is not a fable" in the sense Derrida may have meant.[16] But Cohen's choice to supplant the nuclear with climate may miss an opportunity to supplement it. The "fabulous" nuclear war of the Cold War was not a fable either, given the "nonlocalizable" nuclear war represented by many real detonations, all of which had material effects. Replacing the nuclear with climate as the paradigm for criticism risks perpetuating the silences in those earlier Cold War fables—silences that have ongoing effects on nuclear policy. If, in the first nuclear age, radiation exposure for uranium miners or downwinders, Pacific islanders or Nevadans or Kazakhs, was an acceptable risk in the face of the greater danger posed by the Cold War, in the second nuclear age, it has become, unfortunately, the "war on global warming" (as the cover story for *Time* magazine dubbed it in 2008)[17] that has served to rationalize radiation risks, as, in the face of climate change, we need, or so the story goes, to learn to "stop worrying and love nuclear power."[18] Reading, then, against the grain of the "changing climate" of nuclear energy, a rhetorical cleansing of the technology in the context of peak oil and a warming world, I would like to set Fukushima back into the larger context of the first nuclear age, a contextualizing that challenges the kind of "rational" risk/benefit analyses that have accompanied its aftermath.

For those who have long campaigned against nuclear power, the Fukushima incident that began in March 2011 offered a sense of déjà vu, and one only heightened when, in April, the International Atomic Energy Agency (IAEA) upgraded the accident to a level 7 on the INES (International Nuclear Event Scale), putting it at the same magnitude as the Chernobyl accident.[19] Commentators were quick to argue that this was misleading, that evacuation policies were more efficient, that the Fukushima plant had released far less radiation into the environment, but in the case of a disaster that was (and, considering the persistence of the toxin, is) likely still ongoing, one wonders what the future will say of the comparison. What will the death toll have been of this disaster, and when will we know? How far will the radiation spread and how long until it decays? And, since radiation can neither be seen nor felt, whose analysis of the matter will we trust? As Japanese officials went on camera to eat Fukushima produce, they bore an uncanny resemblance to the British agriculture secretary who, at the height of the BSE (mad cow) scare, fed his young daughter a hamburger to reassure the public of beef's safety.[20] Will such moves have been benign PR stunts or a reckless endangerment of human life? Watching the cautiously optimistic face of the Japanese chief cabinet minister as he ate a Fukushima strawberry, one cannot help but share the sentiment voiced by one of the Fukushima residents, quoted in another context by the CBC: "I don't think they are telling us the truth. Maybe even they don't know."[21]

In the case of Japan, of course, nuclear disaster cannot help but conjure images of that earlier nuclear disaster at the close of World War II, though proponents of nuclear power have naturally tried to separate the warring and peaceful atoms. And as a U.S. citizen residing in Canada, I find myself wanting to triangulate the responsibilities conjured by the uranium cycle. If Canada's tar sands and Keystone XL pipeline have highlighted the country's role as resource colony for the world, this is hardly a recent phenomenon. Alongside our "terror-free" oil, Canada was until very recently first in the world for uranium production, and, as few inside or outside of North America seem to know, Canada supplied uranium for both Japanese nuclear catastrophes. TEPCO (Tokyo Electric Power Company), which operated the Fukushima plant, was one of many customers of Cameco, one of the world's top uranium producers with mines in northern Saskatchewan, and the U.S. bombs that were dropped on Hiroshima and Nagasaki were fueled by ore mined from the Northwest Territories—by Eldorado Limited, one of the Crown corporations that merged to form Cameco in 1988. My point here is not to conflate these events, but rather, in the spirit of the Chernobyl comparison suggested by the coincidence in INES number, to suggest the importance of thinking across time and space. Fukushima is not an isolated disaster that flashes up as an aberrant event that passes with time—or rather it is that, but it is also a slow-moving catastrophe that renders resolution nearly impossible. The relationships and responsibilities revealed and newly generated by such disasters reach around the world, back through history, and far into the future.

Fukushima is a disaster of the second nuclear age, a historical moment in which climate change seems to have trumped nuclear fears, in which a meltdown is presented as an acceptable risk for lower carbon emissions. But to view Fukushima *only* in the light of the second nuclear age is to replicate a larger phenomenon in the staging of risk, a sleight of hand that narrows the data to produce desired results. As Beck notes, in the case, for example, of Chernobyl, statistics regarding casualties range from fifty to a million, with the IAEA coming in at a modest four thousand in 2005.[22] The differing tolls depend on how one frames both the disaster (in terms of immediate deaths at the site or of those less directly exposed) and the "casualties" (is mortality the metric? what of those treated for cancer who survive?). Analogously, putting Fukushima into a larger context, one that includes the first nuclear age, may produce a different picture. In a comment in the *Bulletin of the Atomic Scientists*, Charles Perrow, a sociologist whose work on accidents has long been a part of the risk canon, notes an uncanny resemblance in the rhetoric of harm in radiation exposure across historical catastrophes:

> The denial that Fukushima has any significant health impacts echoes the denials of the atomic bomb effects in 1945; the secrecy surrounding Windscale and Chelyabinsk; the studies suggesting that the fallout from Three Mile Island was, in fact, serious; and the multiple denials regarding Chernobyl (that it happened, that it was serious, and that it is still serious)."[23]

In the rhetorical context of climate change, of course, "denial" takes on new meaning. And Perrow is indeed borrowing power from that charge, suggesting that nuclear denial ought to have the same signification as climate denial, both seen as unscientific and dangerous. In each instance, Perrow notes, downplaying risks led to specific results, for "nuclear denial creates scientific ambiguity that provides cover for governmental and commercial interests and allows nuclear power to continue expanding worldwide."[24]

Perrow usefully bridges weapons and reactors, the first and second nuclear ages, but his list of denials could easily be broadened. Magic as it appears to be, nuclear power does not originate at the plant, nor do the hazards associated with it. Similar sorts of radiation denial have accompanied all moments of the uranium cycle, from mining to disposal, producing and reproducing, as Jacob Darwin Hamblin notes, "the motifs of nuclear history": the "risk society motif," cast, in his essay, in terms of the "media discussions of risk society," "which tend to emphasize that we live in a risky place, that we take a risk every time we drive a car";[25] the "watchdog motif," the tendency to believe that the IAEA is a watchdog, when in fact it exists to promote nuclear energy; and the "nuclear fear motif," which casts concerns about radiation as irrational and emotional. Each of the motifs that Hamblin highlights has the effect, of course, of minimizing (and particularly in the last case, feminizing) risk. As Hamblin's choice of the word "motif" might remind us, these are narratives, and, given the difficulties in detecting and assessing the dangers of radiation, given that laypeople generally cannot access its "truth," narratives are in some ways all we have. Such narratives must be read critically, even if we, laypeople, are armed only with doubt and the ironies of history.

These repeating motifs represent ahistorical constants that bridge first and second nuclear ages, a kind of amnesiac repetition that requires a continual corrective historical situating and contextualizing, the type of work practiced by scholars of risk society, who, Hamblin notes, point to the "organized irresponsibility" of the system and offer "pleas for accountability in an era when predictable mishaps are too easily marveled at as unpredictable perfect storms."[26] Hamblin concludes his essay by informing readers that the IAEA "does not claim responsibility for the actions of the entire nuclear industry. No one does. No one ever has."[27] This ominous reminder of the abdication of responsibility in the face of a technology that has the potential to render large areas of the planet sacrifice zones offers a further challenge for risk critics, those who might, against the grain of the "organized irresponsibility" of Fukushima in particular or risk society in general—the claims that TEPCO could not have predicted the "perfect storm," that Cameco bears no responsibility for supplying the fuel, or GE for designing the reactors, or the Japanese government for deregulating the industry, or the IAEA for promoting it—experiment with something like a discursive "organizing of responsibility," tracking the atom along its routes and connecting historical legacies to present practices.

Climate change requires new theoretical frames and approaches, not only to confront something wholly and radically unprecedented, but also, as Tom Cohen suggests, to "reread ... the archive in its entirety with different referentials."[28] And so, in the spirit of Žižek's description of revolutionary politics, "a movement of *repeating the beginning* again and again," his suggestion that "everything should be rethought, beginning from the zero-point,"[29] I would like here to go back to the beginning of the first atomic age, the moment with which I experimented as an origin of global risk in chapter 1, but this time I would like to track that always elusive "zero-point" back further, not to the Trinity explosion and Oppenheimer's iconic hubris, not to Fermi's ironic wager, or even to his unshielded atomic pile, but to the origins of those materials in a remote mine shaft on the shores of Great Bear Lake in Canada's Northwest Territories, for though this story might seem a closed book, a local moment in Canadian history, in the twisting and open-ended narrative of risk society, both beginnings and endings are elusive.

BACK TO THE BEGINNING

When Eugene Rabinowitch, in his postmortem on the scientists' movement, "Five Years After," posited the twin origins of the movement as the atomic pile and the Trinity test, he was doubtless correct—these were the moments that gave the scientists "a vision of terrible clarity: They saw the cities of the world, including their own, falling into dust and going up in flames."[30] But what the scientists likely did not see was the origin of the material for these crystal balls, the pitchblende ore mined, transported, and refined as the matter they would transmute into global risks. Though there are reports that Fermi's group visited the processing center in Port Hope, Ontario, there is no evidence that they made the additional arduous 2,100-kilometer journey back to the source.[31] That they may not have calculated the inhabitants of the community at what came to be called Port Radium as a part of this "terrible vision" is hardly unique, of course. Mining is seldom considered in conversations about nuclear energy, whether for warheads or reactors. Indeed, those at the mine site also knew nothing of the scientists—certainly, like all world citizens, they would have been ignorant of the secretive Manhattan Project, but, given the remoteness of Great Bear Lake, they also had little idea of the nature of the war effort in general. Nonetheless, their contribution—and, as it turned out, personal and collective sacrifice—was integral to the atomic age. Revelations about these relationships, their history but also their legacies, have prompted local residents, scholars, activists, and documentarians alike to consider the ecological, historical, cultural, and ethico-political implications. The three texts I will consider here—Peter Blow's documentary *Village of Widows* (1999); Marie Clements's play *Burning Vision* (2003); and David Henningson's documentary (an attempt to update Blow's) *Somba Ke: The Money Place* (2006)—all return to Port Radium in order to intervene directly in the "motifs" of nuclear history, recasting risks as uneven and unacceptable, revealing the paucity of "watchdogs," and honoring the emotional and physical scars that the rational fear and material devastation of radiation brings.

Each of these texts tells, in effect, the same story, of how outsiders from the South came to Déline, the community near the lake, to mine for uranium, employing the native Sahtu Dene people mainly as ore carriers along the network of boating routes and portages that linked the remote mine at Port Radium to the railway line that would take the unprocessed ore to be refined in Port Hope.[32] Each story then traces the lack of information and regulation, the exposures to radiation, and the tragic legacy in cancer rates in the area, especially among men who mined and carried the ore and the children who would have been especially vulnerable to the effects of exposure. And each situates this ongoing story in the context of what Gayatri Spivak once called the "vanishing present," with each iteration subsequently dated as the story proceeds into the incalculable future.[33] The story of Port Radium is the story of how a small, seemingly very isolated community is inextricably connected to global transformations, its uranium marshaled first for the war on fascism and *perhaps* later for the war on global warming. That "perhaps" marks the openness of the question, however, both for those on site and for those of us elsewhere, energy consumers in a warming world.

In his more recent version of the story in his book *Highway of the Atom* (2010), Peter van Wyck offers the paradigm-shifting suggestion that what happened at Port Radium "amounts to a paradigm . . . of the workings of ecological disaster. Something happens here and now, because it really happened over there and then."[34] Cancer happens in Déline because uranium transport happened before, an activity that only became toxic retroactively when the effects become visible. Van Wyck's statement is itself paradigmatic of a risk criticism that, in the interest of precaution and environmental justice, posits certainty in the face of uncertainty, apparent in his rhetorical bridging of the events, the "happening" and the "really happening" as well as the causal "because." Narrating cause and effect, here, seems to require a kind of documentary approach, a staging of those events that renders them real, not just materially but discursively, which, in the logic of policy amounts to the same thing. But the choice of the word "paradigm" suggests that this example must also be mobile, situated in time and place, in the real, but also rendered abstract such that it might be ported from one example to another. Indeed, we might extend its paradigmatic status, for the ways in which this slow catastrophe has been told, widening or narrowing in implication, keyed to new presents and imagined futures, also offers something of a paradigm, this time of an iterative practice of retracing the past in order to grapple with changing presents. As each iteration of the story of Port Radium, of what happened then and what is happening now, moves into a different present and future, its signification changes, from the focus on restitution in *Village of Widows* to the new era of uranium mining in *Somba Ke*. Each repeats the beginning, in other words, but to a slightly different end, the sphere of relevance in each iteration widening such that the event simply cannot be contained.

Telling the story of uranium, in which causes and effects are distant in space and time, in which responsibilities span the globe, presents a representational challenge that Rob Nixon associates with slow violence more generally: "From a narrative perspective, such invisible, mutagenic theater is slow paced and open ended, eluding the tidy closure, the containment, imposed by the visual orthodoxies of victory and defeat."[35] Nuclear denial—the suggestion that low levels of radiation have no harmful effects—persists in large part because of the difficulties of drawing causal connections in such circumstances, both because of the time lag between cause and effect, given the long latency period of cancer, and because carcinogenic substances are so plentiful that a single culprit can be difficult to identify, precisely the alibis that enable the organized irresponsibility of risk society, in which no one is responsible (or liable) for harms that may come to pass. Beck cites the example of a lead crystal factory in Altenstadt, which released pollutants like lead, arsenic, and fluoride vapors, affecting the health of human beings and the environment in the vicinity. Given, however, that there were several factories in the area releasing similar pollutants, even in the face of clear evidence of harm, no responsibility could be ascribed. Scaling this up or down makes this example paradigmatic: From local producers of toxic chemicals or radiation to CO_2 producers worldwide, "It is becoming impossible to ascribe harms suffered by many people—and at the extreme by everyone—to an author in conformity with valid legal norms and to assign responsibility."[36] In the case of the Dene ore carriers and white miners at Port Radium, paradoxically, the harms suffered are not by too many but by too few. Though there may seem to be elevated cancer rates among those in Déline, the population is so small as to frustrate the normal protocols of epidemiological study. Records of Dene workers are scarce, given the informal economy that likely prevailed, and accounts of the handling of uranium along the route largely anecdotal. In this context, in which the very history of the event is in question, assessing the damage and apportioning the liability, at least in the usual actuarial fashion, becomes a difficult, if not impossible, task.

Peter Blow nicely captures the complexities of cause and effect, beginning and end, in the evocative opening sequence of *Village of Widows*.[37] The film begins with a funeral, the scenes desolate, wintry, despair visible on the mourners' faces as they move toward the grave. Voice-overs from the narrator and then from two women interviewed, Shirley Baton and Cecille Baton, inform viewers of the high rates of cancer in Déline. As the scenes of mourning progress, a cappella singing replaces the voices, a slow sort of funeral dirge appropriate to the scene, but as the melody descends toward what sounds like a resting point, its final note is replaced by an atomic explosion, which accompanies the cut to archival footage of the bomb, jarring the viewer from his or her contemplation of the scene. For the uninitiated, as most would have been at the time of the film's release, such juxtaposition would necessarily raise questions about the relationships among these events—the death, the cancer, and the bomb. Though these images and sounds have the effect of revealing fact, the imagistic vehicles are somewhat taken out of context. The explosion is clearly not at Hiroshima; the funeral may not be that of an ore carrier.[38] The images are not necessarily, then, literally connected; rather, it is the effect of the juxtaposition that is to reveal the truth. In a case in which evidence is scarce, archives closed, histories unwritten, dangers occluded, and responsibilities evaded, the story must be as much made as uncovered, or, as Peter van Wyck puts it, "In a way, until Blow's film was made, there was no story to be told."[39]

Village of Widows thus works to establish cause and effect, documenting two very different attempts by Dene people to establish responsibility. The first was simply an appeal to the government of Canada, which had operated the mine from 1942 until its closure in 1960, to acknowledge the toxic legacies of uranium and to clean up the site; this type of accounting required historical evidence of the Dene's direct involvement and likely exposure as well as accurate testing of present dangers, the levels of radioactivity in the tailings on the lake bottom and on the site of the former mine. The second form of responsibility is, however, somewhat surprisingly, that of the Dene themselves. Here, there was no need to establish intent to harm, no need to trace the bureaucratic lines of liability and accountability—the supply chains, the processing centers, the governments responsible for organizing the systems, the scientists and technicians who developed the bomb, the military who used it. The discovery that the ore mined in their territories fueled the weapons used in Japan was, alone, enough to evoke a sense of moral responsibility. As van Wyck describes it, Dene community members "disavowed the vortex of history and archive, and the also vortical administrative discussions of sovereignty, rights and self-government, and moved directly into the realm of the ethical," by themselves journeying to Hiroshima to offer an apology.[40] Over the course of the film, then, Blow tracks two journeys made by the Déline Uranium Committee, one to Ottawa and the other to Japan, chronicling the very different results of each.

Blow's documentary thus has several burdens—to document these acts of responsibility, and also to chronicle the narrative to which these acts respond, to render visible the history that would enable the viewer to contextualize those journeys in the present. Framed by the two attempts to find justice is a fairly conventional historical narrative, pieced together from archival photographs and footage and interviews with residents, journalists, and academics. The spaces in the story are spanned by an authoritative voice-over that carries the history forward. This story begins, in effect, before the beginning, by referencing a prophecy (repeated in each of the texts I consider here) that the Dene believe foretold the fate of the mine, the uranium, the Japanese, and the Dene themselves. This prophecy, described in many accounts, including George Blondin's collection *When the World Was New* (1990), predicted that, as Elizabeth Kordican explains in the film, "in the future, when that mine closed down, things were going to change for the worse. He said that there would be many hardships because of those rocks, and the people are going to suffer." *Village of Widows* then tracks the history of the mine at Port Radium, from its staking by prospector Gilbert LaBine[41] in 1930, to its development, first for radium and then, when the war effort demanded it, uranium in the 1940s—and the consequent shift of the mine from private to government control. Alongside the official history, Blow presents the stories of Dene ore carriers and white miners and their families, highlighting their daily interaction with the ore, as it spilled on them or into the lake, as they made tents out of old ore sacks or played in sandboxes filled with tailings, as it found its way into their clothes, food, and water. The consequences of this exposure come in the litanies of cancer deaths, whether in Shirley Baton's "My dad died of cancer; my aunt died of cancer; my grandmother died of cancer; my mom is suffering because of her sickness," or in Déline elder George Blondin's "I had seven children; four of them died plus my wife, and they all died of cancer."

If the Déline Uranium Committee's visit to Japan offers a model for ethical response to uranium, the film also chronicles abdication of such responsibility. The committee is invited to send a delegation to Ottawa to meet with representatives from the federal government, only to find that, at the last minute, plane tickets that had been promised weren't delivered, meetings scheduled were confused. Though a meeting is ultimately held, the narrator reports that, "for all the expressions of sympathy, in the six months since the Ottawa meeting, there has been very little progress on any of the community's fourteen points." Whether any government officials were approached for more active participation in the film is unclear, but they certainly do not appear on camera, leaving the viewer in the position of the Dene, shut out of whatever official machinations might have occurred behind the silence and inaction. In hindsight, we know that the government did act, producing a report that Henningson highlights in his later film, but at the point at which *Village of Widows* premiered, this future was still to come.

If, as Jill Godmilow points out, "audiences seek and expect closure, even in documentary films,"[42] this is hard to achieve in the case of a documentary chronicling a substance likely to be toxic for, Blow's voice-over informs us, the next eight hundred thousand years. As a result, the film plays with beginnings and endings, closure and containment, producing, in the process, both its own forms of narrative closure and a particular viewing subject. Opening with the funeral and the bomb, the film, in effect, begins at the "end," at least historically, in Déline, which, the voice-over informs us, is known as the "village of widows" as a result of the deaths of ore carriers, juxtaposing that to the "beginning" of an earlier event in Hiroshima, Japan. But as *Village of Widows* proceeds, ends and beginnings get confused, less tied to chronology. Port Radium is, after all, also the beginning, insofar as it was the source of uranium, the end result of which was the bomb. But as the present scenes of mourning suggest, that explosion was hardly the end. Thus, when one of the enthusiastic participants in the Dene journey to Japan connects "these two groups of people, the first ore carriers of the uranium that went into the bomb with the first people on whom the bomb was dropped; the beginning is meeting the end," the tidiness of his assertion, which comes toward the beginning of the film, is undercut by the evidence of ongoing harm.

The story, of course, cannot be over, but the film must end, and *Village of Widows* evinces a desire for some sort of satisfying closure. Though one might have hoped for a government response that would match the ethical commitment of the Déline Uranium Committee, the film concludes with a different form of resolution—a reinvestment in traditional values in Déline, which might sustain the spirit of people even in the face of official inaction, and the visit to Japan, as committee members participate in the lantern ceremony in Hiroshima, sending messages afloat to "console the spirits of the dead." The film as a whole has this commemorative spirit, as the residents of Déline and Hiroshima come together to acknowledge those ties that had not only not been told, but had been structurally silenced.

In its chronicling of the Dene in Japan, the film suggests the power of subpolitics—that an international grassroots politics of "being at risk" might thrive even in the face of government inaction. What is less clear,

168 RISK CRITICISM

5. From *Village of Widows*. (Courtesy of Peter Blow, Lindum Films.)

however, in this resolution, is the place of the viewer him- or herself. The intimacy of the camera produces the audience as a kind of privileged voyeur, along for the trip to Hiroshima, or watching the Dene dancers in the final scenes. Thus *Village of Widows* may risk offering something of what Godmilow calls the "classic contract arrangements that the documentary film usually proffers":

> The audience is invited to believe: "I learn from this film because I care about the issues and people involved and want to understand them better; therefore, I am a compassionate member of society, not part of the problem described, but part of the solution." The documentary film knits us into a community of "we"—a special community by dint of our new knowledge and compassion.
> The real contract, the more hidden one, enables the viewer to feel: "thank God that's not me."[43]

The "not me" in this case seems to refer to both sides of the issue: we are not Dene, for we require the education the film provides, but we are also not the shadowy and unresponsive government. We certainly leave the film feeling a part of a community, then, but whether we feel a sense of our complicity as well as a sense of compassion is an open question. The narrative closure of the film—the ending returning to the beginning, the Dene in

Japan; a final turn to the resilience of tradition—may very well function, as Godmilow suggests closure does, to "send the audience out of the theater (and/or off to bed) feeling complete, whole, and untroubled."[44] Or perhaps we are to be troubled in spite of the end, for given the open-endedness of the hazards, the resolution of the film feels almost artificial.

BEGINNING AGAIN: TROUBLING NARRATIVE

Covering much the same terrain, Marie Clements's play *Burning Vision* (2003) reads in some ways as an adaptation of Blow's film for the stage. Indeed, Clements credits Blow and several of those interviewed for his film in her acknowledgments. "Complete, whole, and untroubled" is precisely not, however, the affect likely produced by the play. *Burning Vision* opens with a timeline that stretches across a map of Port Radium and Great Bear Lake, offering the promise of resolution, as, at one end, a prophet (the "Dene Seeer") in the 1880s foresees the development of the bomb, and, at the other end, in 1998, Déline residents visit Japan to pay their respects, followed soon thereafter by the debut of *Burning Vision* in 2002. If this timeline would seem to suggest closure, with the Déline residents accepting responsibility for their unwitting complicity, and, presumably, audience members, educated by the play, doing the same, the play itself offers no such comfort. Those familiar coordinates of time and space—the timeline and the map—offer an orienting guide that the play disrupts entirely, as time and space are stretched and compressed to accommodate the disparate speeds and dispersed geographies of the violence of the uranium cycle.

If, as van Wyck suggests, the story of Port Radium is paradigmatic of ecological disaster because "something happens here and now, because it really happened over there and then," *Burning Vision* collapses these times and spaces, making the narrative jarring, difficult to follow and to understand—as Theresa May asks, "How do we talk about the 'world of a play' in which conventional boundaries of time and space evaporate, and different historical moments overlap in a kind of double and triple exposure?"[45] The result, as May's reference to "double and triple exposure" suggests, is a play almost cinematic in its effects, a quality particularly marked in the stage directions, which, as in the opening "intense darkness pierced by light revealing warning scenes of the human noise of pain, grief, loss, and isolation," seem to invite surround sound and digital animation. But here the differences in time and place are no longer differentiated by those familiar conventions of documentary film, in which the qualities of the photos and footage—grainy black-and-white versus high-resolution color—might allow us to contextualize events in historical time.

As in Blow's film, *Burning Vision* opens with a juxtaposition of the ends and beginnings of the uranium cycle in the bomb and mining. Here, the detonation is clearly a test, as the voice-over, a U.S. radio announcer, informs us, there are observers "adjusting their goggles" in preparation for the blast. The countdown then culminates in the "sound of a long, far-reaching explosion that explodes over a long, far-reaching time," an image that captures the geographical and temporal reaches of the uranium cycle, both forward and back.[46] This test may well be Trinity, suggesting this as origin, but this narrative is quickly shattered, as the "long time" ends up stretching backward as well as forward, as the flash of the bomb is then replaced by a darkness subsequently pierced by "two flashlight beams," representing the uranium prospectors, named in the collective "Bros. La-Bine," who set the cycle in motion.[47] As these lights swing around the stage, they illuminate other characters, making them visible to the audience though not to the prospectors themselves. Over the course of the play, the characters become aware, first of the objects on the stage, those global commodities that tie them all together, and only very gradually of the human beings behind them.

Thus, as in *Village of Widows*, the beginning meets the end, the bomb is juxtaposed to the prospecting, but *Burning Vision* takes this juxtaposition a step further, imagining all of the "then and theres" and "now and heres" together simultaneously on the stage. Appearing alongside Bros. LaBine 1 and 2 are a boy representing uranium, a Dene ore carrier and his widow, a grandmother and grandson from Japan, and an American bomb test dummy aptly named "Fat Man," a radium dial painter, two stevedores, a miner, and a Métis baker whose father runs the Hudson's Bay Company store in the area. These players are all laypeople, not the experts responsible for assessing risk, but the recipients of the risky materials who are reliant on experts for the interpretation of hazard. Though experts are invoked by the characters, they are not in evidence. Rather, Clements represents the staging of risk and expertise in the media by incorporating radio personalities and voice-overs from the period, including Iva Toguri, or Tokyo Rose, a Japanese American radio personality, and Lorne Greene, or the Voice of Doom, the announcer for the CBC's *National Bulletin*. Also included are anonymous radio announcers whose voice-overs, in Slavey, Japanese, and English, conclude the play. Though these voices provide information and a sense of community, they do not do much to illuminate the risks that the characters face. As Fat Man, the bomb test dummy, puts it, "You see anybody from the government sitting here. No, they're the ones pushing the buttons and we're the ones sitting in our living rooms watching the goddamned news."[48]

Burning Vision thus stages sight and insight, ignorance, revelation, and knowledge, depicting the impossible education of the characters, even as the play itself has a pedagogical function for the audience. The "burning vision" of the title refers at least in part to the prophecy of a Dene See-er (the same prophecy to which Blow's film alludes) who, in the past that is present in the play, predicts that a rock mined near Great Bear Lake will be dropped on people who look like Dene—clearly, a reference to the Japanese, but also, given the slow violence also depicted, to the Dene themselves. Clements reveals this prophecy even as it comes to pass, thus demonstrating the ways in which risks, which are largely prospective, turn into catastrophes when risk management fails. A staging of this side of what Beck calls the "becoming-real" of risk,[49] its passage from projected risk to present catastrophe, comes in the simultaneous appearance of the brothers LaBine and the widow, who represents the future from the prospectors' perspective. Her husband's death is a result of carrying the radioactive ore that they hope to find. Clements's placement of these characters performs, in effect, a kind of future anterior, showing that the mining will have been a cause of cancer, and though the prospectors cannot initially see this, the audience clearly can. The play thus stages relationships of cause and effect, suggested but (still) not proven.

This technique of simultaneity, of overlapping times and spaces, reaches its apex in the final movement of the play, which is syncopated by the rhythm produced by the repetition of three sounds: the tick, tick, tick of a clock associated with the radium dial painter; the click, click, click of Geiger counters, associated with the prospectors; and the beat, beat, beat of a human heart, presumably that of the child conceived in the union of Rose, the Métis baker, and Koji, the Japanese bomb victim who has been magically transported to Great Bear Lake by the explosion. Representing the past, present, and future, these figures highlight questions of temporality and responsibility. One of the brothers LaBine, frustrated with the ways in which critics are already calling his practices into question, wonders: "Christ, how many discoverers have to listen to this bullshit. I'm trying to run a corporation here. I'm trying to keep men employed so they can feed their families"[50]—familiar narratives that accompany dirty industries today. Immediately following this passage, Rose, who still is not interacting directly with the brothers LaBine on stage, says: "You have to believe in the future no matter what."[51] While this could be read as hopeful—that is, she is concerned about the future of her unborn child—it can also be read as a response to the time scale on which the prospector's thinking relies, which privileges an immediate present over an extended future.

As LaBine Bros. 1's comment suggests, even if we cannot blame the prospectors for lacking the prophetic sight of the Dene See-er, it is not exactly true that cancer deaths from uranium were unpredictable. To emphasize this, Clements also includes the radium dial painter, who, like the early twentieth-century historical figures she represents, periodically licks her glowing paintbrush. Her eventual disfigurement over the course of the play—the stage directions in the final movement read: "*Half her face is missing and her beautiful hair is entirely gone*"[52]—highlights the known hazards of the material. This moment comes, in fact, in the midst of an extended rumination again on the part of one of the Bros. LaBine on issues of expertise, regulation, and knowledge. He begins: "Jesus, what a beautiful sound . . . click . . . click . . . click . . . As if they could send a bunch of so called 'experts' up here and convince me that this uranium is like a goddamned grenade going off."[53] A few pages later: "Christ, these cry baby 'experts' say it's dangerous to even breathe a few little particles in, never mind handling it, but . . . what does anybody really know."[54] And finally later: "The hell with you, and the HELL with them. The government knows what it's doing and the government is behind me."[55] As these rationalizations continue, the brothers suddenly become aware of the characters that have previously been hidden by time and space. The stage directions read:

> *The sound of the Geiger counter gets closer and closer and it is now louder than it's ever been as it clicks toward* ROSE, *circling in on her.* BROS. LABINE 1 *looks in horror as the Geiger counter hits her belly . . .*
>
> *The sound of a bomb falling as* BROS. LABINE 1 *looks at* ROSE *and then directly at her belly.*
>
> BROS. LABINE 1: I didn't know.[56]

Clearly, knowledge is here contested. The prospector knows and he doesn't know; experts provide controversial reports; governments charged with regulation fail to regulate. Some of his ignorance is willful and some structural. Waiting for the future, represented by Rose's unborn child, to reveal the hazard will have been too late.

As it turns out, the "horror" of revelation, a moment that implies direct knowledge that harm will have come to the child, is misleading. In the end, the child survives and is apparently unharmed, though his parents, Koji and Rose, have passed away, inexplicably offstage. What the play offers is thus not the certain knowledge of risk management but the uncertainty of the precautionary principle, a different orientation toward the future, with different implications for thinking about knowledge and responsibility. As François Ewald notes, "Under the old approach to responsibility uncertainty of knowledge was innocence";[57] thus, lack of scientific certainty of harm meant lack of responsibility when the harm came to pass. Bros. LaBine 1's "I didn't know"—meaning 'I didn't know with certainty, in a way that was uncontested'—would thus have relieved him of responsibility. In the precautionary regime, however, responsibility shifts and one is judged, Ewald says, "not only by what one should know but also by what one should have or might have suspected."[58] Precaution thus "invites one to anticipate what one does not yet know, to take into account doubtful hypotheses and simple suspicions. It invites one to take the most far-fetched forecasts seriously, predictions by prophets, whether true or false."[59] If, at the start of the play, Bros. LaBine 1 scoffed at the idea that a rock might be dangerous—"HERE COMES THE BLACK ROCK . . . Boo!"[60]—by the end, he comes to see that the prophetic "Indian fairy tale"[61] was a form of knowledge, the Dene See-er's vision a "prediction by prophets" that ought to have been taken seriously.

At one point in *Burning Vision*, one of the Bros. LaBine says, "Indians are always telling stories. Trouble is they don't know when a story should end, and reality should begin."[62] In the "becoming real" of world risk society, though, this distinction between story and reality is decidedly blurred. Countering the Bros. LaBine abdication of responsibility, *Burning Vision* suggests that partial knowledge ought to activate the precautionary principle. When Clements's Dene See-er, an expert of a sort, perhaps, but certainly not an atomic scientist, describes his vision, he recalls, "I saw a flying bird, big. It landed and they loaded it with things. It didn't look like it could harm anybody, but it made a lot of noise. I watched them digging something out of a hole in the earth and I watched them rise [sic] it to the cool sky until it disappeared and reappeared. Burning."[63] "I wondered . . . if it would harm our people," he continues. "The people they dropped this burning on . . . looked like us, like Dene."[64] His suspicion, clearly, is that it will cause harm, despite his relative ignorance of airplanes or nuclear weapons technology. And as we, laypeople, confront the potential dangers of world risk society, armed with a similar sort of nontechnical vocabulary, we too might hazard prophecies. Indeed, Clements's See-er seems to call us to just this sort of responsibility when, at the opening of the Third Movement, he turns from first to second person, asking, "Can you read the air? The face of the water? Can you look through time and see the future? Can you hear through the walls of the world?"[65]

Burning Vision thus tracks a kind of revelation, but the play does not resolve neatly in this new knowledge. Indeed, Clements seems self-consciously to play with the literary conventions that tend, in other narratives, to give closure. Noting the ways in which the characters far apart geographically and culturally are brought together in the space of the theater, Theresa May has suggested that *Burning Vision* stages a "transnational neighborhood."[66] In fact, at one point, as bombs fall on the test site, Fat Man says, "I think they got my neighbour down the road. If they got my neighbour, I'm next, then you're next, that's how neighborhoods work"[67]—here, as in Beck's risk society, community comes as a result of being at risk together. But the play also raises questions about the feasibility of this sort of risk cosmopolitanism, in part by exaggerating the function of intimacy through a parodic marriage plot. Koji, the Japanese character whose appearance in Great Bear Lake inverts the Dene visit to the Hibakusha in Japan, becomes romantically involved with the Métis breadmaker, Rose, and they conceive a child. Other characters are similarly united, with Fat Man paired with Tokyo Rose and the boy who represents uranium, and the uranium miner awkwardly wooing the radium dial painter—mini nuclear families in the face of nuclear threat. These romances and quasi-marriages lend something of the resolution of comedy to what is otherwise a tragic story, but the play also evinces an ironic awareness of the dangers of too easy resolution. When Rose asks Koji, "If you make me yours do we make a world with no enemies?"[68] her own parents' experience suggest a less than utopian answer. Koji echoes and inverts Rose's question: "If we make a world, we will make one where there are no enemies?"[69] Though this reads as a statement, it is concluded with a complicating question mark rather than a confident period. And if these subtle complications of the marriage plot are all too subtle, Fat Man, the character tasked with (often crass) frankness, says of his own relationship with Tokyo Rose, "It just shows different cultures can get along if we're all willing to sit down and fuck ... Talk."[70]

The play's ending does connect disparate strands of the narrative together, but here too closure is fraught. The final image, described in the stage directions, draws together the caribou and cherry blossoms of Canada and Japan into a single symbolic visual: "*Glowing herds of caribou move in unison over the vast empty landscape as cherry blossoms fall till they fill the stage.*"[71] Here, acknowledging that the release of radioactivity cannot be undone, Clements nonetheless offers an image of life going on, as the glowing caribou continue their migrations. The final words of the play belong to "Koji the grandson," the child of the now-deceased Koji and Rose, who has been adopted by the Widow: "They hear us, and they are talking back in hope over time."[72] The "they" most immediately are his parents, but given the chorus of "announcer" voices, Slavey, Japanese, and Canadian, who precede this statement, "they" are also other voices of the past, reminding us, in these present conditions, that we must anticipate the future, or, as Round Rose puts it, in an extended meditation on the public apology, "be sorry before you have to say you are sorry."[73] There is certainly a closure in this conclusion, but the complications of the play's temporality also render any "conclusion" suspect. Theresa May suggests that *Burning Vision*'s plot "transpires in the split second between that first flash of light and its reign/rain of sudden death," a reading that challenges the very concept of duration.[74] But even this may suggest too much neatness to the beginnings and endings—the opening detonation is "far-reaching"; how far is an open question.

BEGINNING AGAIN, AGAIN: RECALCULATIONS

In the production notes for his documentary *Somba Ke: The Money Place* (2006), David Henningson describes the origins of the film as a series of false starts. He set out, he tells us, to update the narrative that Peter Blow had traced in *Village of Widows*, asking: "What had happened to the Dene and their cause? Did the Canadian government investigate the contamination, and if so, what were the results?"[75] Soon thereafter, he encountered Marie Clements's play, and his plan, pitched to and accepted by Canada's Global National TV, was, at the time of the sixtieth anniversary commemorations of Hiroshima, to have "several Dene and Marie Clements go to Hiroshima in an exchange that would see several Hibakusha (Japanese atomic bomb survivors) returning to Déline." Nicely invoking both earlier tellings of the story, this new film was to repeat those earlier journeys with a difference. As it turned out, however, this plan did not come to fruition. Henningson reports that the Dene he had invited ended up declining; those who did agree to go to Japan refused to be on camera. He filmed Marie Clements reading from her play in Japan, but these scenes did not make the cut in the final version. Most tellingly, he found that "the atmosphere in Déline had changed."[76] The film that was produced is an assertion about what might have changed that atmosphere, a transformation linked to the government's investigation of the contamination as well as to that larger context—the changing climate that recast nuclear energy in such a way that it required yet another return to the beginning, for in the early years of the twenty-first century, as the nuclear industry touted a "renaissance" in nuclear power,[77] interest in uranium mining was rekindling, with Port Radium poised, once again, to be the El Dorado of the north.

It is this rebirth that Henningson describes, returning to the beginning—the uranium mined for the Manhattan Project—but placing it in the context of another return, signaled in the opening voice-over:

> The story of the making of the atomic bomb, the people involved and of its devastating effects have all been well documented. But few realize where in 1942 the Manhattan Project came looking for the Western world's only proven source of uranium, or where, over sixty years later, the worldwide awakening nuclear renaissance is returning for more.

A specter thus haunts Henningson's film, as the history at Port Radium threatens resurrection, not just as past, but as present and future. The film opens, as Blow's does, in Asia, but the focus is now on the desire for power, particularly in Japan and China, where, the voice-over informs us, numerous new reactors are slated to go online to support growing economies. And in this context, the mine at Port Radium, a location where, we learn later in the film, even the waste is "high grade," has returned to the global stage, not as a site of past tragedy, but as a place cleansed and ready for rebirth.[78]

The cast of relevant characters here changes, including some of the same informants that Blow interviewed, George Blondin and other members of the Déline Uranium Committee, looking a bit older now (indeed Blow himself appears at one point), but also those investment bankers, uranium prospectors, and pronuclear activists who are driving the renaissance—and younger community members at Déline who are reconsidering risks and benefits of resource development. This inclusion of nuclear supporters puts the viewer in a somewhat different position vis-à-vis the issues addressed. Though the film is clearly a critique of the nuclear, the interviewees affiliated with that industry seem to believe otherwise. There is little hostility; the implied conversations feel almost conspiratorial, as though we were part of the elite crowd of investors and CEOs who might think "rationally" about risks and rewards. The settings for the interviews are clearly staged to set their subjects in the best light. Cameco's then-CEO, Jerry Grandey, appears comfortable on an overstuffed chair, what appear to be artistic wooden carvings of wolves and other northern inhabitants in the background. Tim Coupland, CEO of Alberta Star, the company most directly involved at Port Radium, is interviewed in a homey setting, side-lit by a lamp, on a couch flanked by a potted plant, an enormous painting with native themes in the background. Given the thrust of the film, these players are not likely to garner our compassion or sympathy, but they may force us to be more conscious of our own complicity in the situation, our own thirst for electricity or financial gain. Henningson interviews Patrick Moore, Greenpeace co-founder and nuclear advocate, who asserts, "My old colleagues in Greenpeace think that we can phase out both fossil fuels and nuclear energy—where are they going to get the electricity? And then they start talking about solar panels and my eyes glaze over." Cut into the interview are scenes of excessive consumption of the sort that characterize any large city at night—the glittering lights, screens, and signs that represent the electricity demand in which the viewer likely participates, at least tangentially.

Updating the narrative in Blow's film, *Somba-Ke* tracks the government's eventual response to the Dene call for justice, for the government did subsequently agree to assess the situation, both by cleaning up any presently dangerous materials and by researching the extent to which the work at or around the mine might have been responsible for the premature deaths of ore carriers. Establishing both present and past dangers was, however, complicated. Radiation hazard depends on how it is measured; epidemiological analyses do not work on very small populations. Henningson's film relentlessly questions the procedures and limitations of the resultant study, including the fact that the "dose reconstruction" portion of the work was farmed out to a consulting agency, SENES, with ties to the nuclear industry. In 2005, the Canada-Déline Uranium Table released its report, which contained the good news that the very minor hazards still present would be remediated (mine shafts closed; surface radioactive materials covered), and, in any case, historic and present exposure to radiation was insufficient to cause higher rates of cancer, a finding that did little to explain the cancer deaths in the area—or to console the surviving families.[79] The recurring motifs of nuclear history that Jacob Darwin Hamblin identifies elsewhere are clearly in evidence in the report, which emphasizes the multiple sources for radioactivity in the north, as well as the multiplicity of carcinogenic risks, some of which, like smoking and alcohol consumption, are taken voluntarily by community members. As in the aftermath at Chernobyl, the report suggests that the main risk that radiation poses at present is psychological; that is, fear of nuclear risk is producing health effects like anxiety and depression (both of which are also carcinogenic). And, in a variation on the "watchdog" motif, the report turns the consultants hired (and the government itself) into watchdogs when in fact they are promoters of nuclear energy. The results of these cleansing motifs are, as Charles Perrow suggests of nuclear denials more generally, instrumental in the expansion of nuclear power—and the uranium extraction that fuels it. Indeed, not long after the report was released, Alberta Star mining company secured the rights to prospect for uranium again at the old site at Port Radium.

As the camera tracks a sea plane landing on Great Bear Lake, we hear the voice of Tim Coupland (brother of author Douglas Coupland, an irony not lost on Henningson), who gives us a précis of events since the Déline Uranium Committee's visit to Ottawa: "We came in after the government spent approximately seven and a half million dollars, and they released a report saying there was no long-term detrimental health effects to the area. I mean all the radionuclides are in equilibrium, and we actually decided to move into the area once that report was released." Describing the economic context in which uranium prices seem to be moving ever upward, Coupland explains: "We started this exercise and I think uranium was at $18, and it's now, as of today it is US$41.80 per pound. And that's pretty significant. We're in what we call a supercycle; it's like a vortex, taking everything upward." The camera then tracks a helicopter, which is likely taking Coupland, his employees, and mineral core samples upward, but leaving a lot on the ground. This motif of aerial views, from and of sea planes or helicopters, dominates in the film, chronicling the ways in which the outside world accesses Déline, "moving in," as Coupland suggests, but likely to "move out" when the mining is no longer profitable. Brought in on Henningson's aerial shots, we too move into and out of Déline, observing events and people as outsiders. Even if we are not, like Coupland, involved in actual extraction, the film seems to suggest that we too might operate with the capacity to hedge our bets on risk, and certainly the desire for energy and enthusiasm for nuclear reactors are not exclusive to developing countries like China.

These views from above give us access to a different, less intimate, perspective on Port Radium, and one with which we, likely outsiders ourselves, are also complicit, making us uncomfortably aware of our outsider/voyeur status. But the film also broadens the frame in other ways, taking the issue south to Navajo (Dine) country, where uranium mining has had similarly devastating effects, and to Australia, where the high environmental price of water usage at mine sites becomes apparent. Such broadening is a corrective, we learn, to the narrowness of the Canada-Déline Uranium Table Report. Henningson interviews journalist Andrew Nikiforuk, whose 1998 series in the *Calgary Herald* first brought the question of radiation exposure at Port Radium to a general audience; Nikiforuk offers an assessment of the 2005 report:

> The final report, which I take it cost close to seven million dollars, is really a travesty on a number of fronts. It's very narrow in its scope. It only addresses the health issues of approximately thirty Dene. It doesn't mention the health problems that the Navajo had in the Southwest in the United States. It ignores the whole fact that thousands of European workers were exposed to dangerous tailings and radon gas in the mine. It doesn't look at any of the broader issues. I mean it is so narrow as almost to be useless. It keeps the whole issue narrowly confined to one place and makes sure that very few Canadians and very few Japanese and very few Americans will ever find out about this history.

As this enumeration of what is not in the report proceeds, the camera itself zeroes in on a mine shaft, the shadow of the helicopter on which the camera is likely mounted also in the frame, offering an image of that narrowing, the view from outside and above. The film then cuts to a shot of the helicopter from the ground, beneath two large canisters that read "Alberta Star," as the voice-over informs us that, in 2006, a second drilling permit was granted on site of Port Radium for the company. "We gambled and we won," says Tim Coupland.

Somba Ke concludes with the voice of the head of the Déline Land and Financial Corporation, Leroy Andre, a man who, our informant Tim Coupland tells us, has the "Alberta sensibility about him" (a phrase that, in Canada, can only signal friendliness to resource extraction). Here, Andre tells of a very different prophecy, not the precautionary prediction of the dangers of the mine, but an eschatological optimism:

> One thing the elders have always come back to is the prophecies, that Déline is going to be one of the last places where people come to and that we would be a very powerful nation. And we truly believe that, so we have to make sure that we can prepare for that, and we can feel those impacts now because of the amount of exploration

180 RISK CRITICISM

that is happening on our land. We have to make sure that we can foresee that and plan for that as a corporation and as a community.

For what people would come, here, seems clear: exploration is the sign; uranium is the source of power, promising a messianic second coming, now to deliver the community to a new era of prosperity and prestige. But what "foreseeing" and "planning for" the future will mean is perhaps less clear here than the "Alberta sensibility" Coupland attributes to Andre. Certainly, the scenes that play as he recounts the prophecy suggest other values in Déline, as the camera shows the lake glittering in the sun (a striking contrast to the Australian tailings ponds seen from satellite photos that the area seems destined to become), children playing on streets, and a fisherman pulling a fish out of a hole in the ice. In an era of environmental devastation, in which freshwater dwindles and temperatures rise, these other assets may, the scenes suggest, offer greater return.

These values are clearly expressed in an article on the future of fishery management in Great Bear Lake, published in the *Reviews of Fish Biology and Fisheries* in 2012, which also opens with one of the prophecies of Dene elder Ayah, a version that echoes the eschatology of Andre's, but here with a slightly different implied source of wealth: "The very last disaster or plague will be 'Famine'[;] it will come from the east and it will affect every living thing. Not even one blade of grass will stand up, every living thing will go ... my people will be the very last ones to have fish."[80] Here, it is not the nonrenewable mineral resources that will provide the wealth for the community, but precisely what that sort of extraction would destroy: the health and vitality of the lake and the fish. Noting that Great Bear Lake is "among the last remaining pristine great lakes of the world," the authors go on to speculate what effects climate change and further resource development (natural gas extraction and pipeline, diamond mining, oil exploration) may have on the area. The spirit of this research, they suggest, is in line with that of the Great Bear Lake Working Group, which, in 2005, expressed its resource management vision: "Great Bear Lake must be kept clean and bountiful for all time."[81]

Invoking the history at Port Radium, Tim Coupland informs us that "a lot of the, you know, standing issues that had never been addressed have been addressed. I think there's closure, and they can move on with that area. What happened there will never happen again." One can easily imagine Clements's ghostly ore carrier, hovering in the background, for such confident calculations, which, especially in the face of the limited liability of the Canada-Déline Uranium Table Report, can only read as ominously ironic. As the iterations of the story at Port Radium suggest, "closure" is precisely what narratives of environmental catastrophe must resist, as the present perpetually rewrites the past in the face of a new imagined future. Given the persistent dangers of the uranium cycle and the detoxifying power of the motifs of nuclear history, the safer bet, *Somba Ke* seems to say, would be to wager against Coupland's prediction.

And yet, as of 2014, Coupland and Alberta Star appear not to have "moved on with that area." The company's website mentions Eldorado and Echo Bay in its "Assay Results from its 2008 Drilling Program," noting the type and concentration of minerals discovered in the initial sampling of these and other properties, but when the news release moves toward the future, the economic downturn seems to have tempered the company's enthusiasm for developing new mines. "In today's turbulent market," it reads, "it is cheaper to buy production than to explore and drill for production and resources."[82] The turbulence here is doubtless the global market, but local ambivalence surrounding uranium mining may also have contributed.[83] Today, with Fukushima tempering the enthusiasm around the world for nuclear power, the "supercycle" of uranium pricing is decidedly subdued. Subsequent postings on the website suggest that Alberta Star has moved into less financially risky—but also ecologically devastating—investments in oil and natural gas, as the climate of nuclear energy shifts again, climate change notwithstanding. Former Cameco CEO Jerry Grandey retired just months after the Fukushima accident. Interviewed by Toronto's *Globe and Mail*, he acknowledged that the industry was entering a "choppy period," but he also confidently predicted that nuclear power will rise again like a phoenix from the reactor's flames: "The phobia about nuclear radiation will cause a pause in some new construction, but—I emphasize—just a pause. Once people reflect on the risks associated with other energy sources, they'll come to the rational conclusion that nuclear, even in the worst circumstances, is still better than the alternatives."[84] Drawing on messaging that Susie O'Brien has noted characterizes the oil industry as well, Grandey insists, in effect, on "TANIA," or "There are no *ideal* alternatives."[85] Such "rational" cost-benefit analysis acknowledges the dangers of nuclear industry, even as it, at least implicitly, acknowledges the possibility of other truly "alternative" sources, no single one of which is likely, in itself, to be "ideal." The ideal is likely something closer to a diverse, decentralized energy system that is sensitive to local conditions, not the kind of development most conducive to the consolidation of money and power that is represented by oil, gas, or nuclear. Invoking a version of the "Nuclear is clean air energy" slogan, Grandey suggests, "If you're truly worried about climate change and air pollution, nuclear has tremendous benefits."[86] This might be a surprise for those downwind of Fukushima Daiichi. As Naomi Klein points out, responding to this sort of "solution" to climate crisis, "Nuclear power and geoengineering are not solutions to the ecological crisis; they are a doubling down on exactly the kind of reckless, short-term thinking that got us into this mess."[87]

Fukushima, many commentators suggested, was an example of those "worst circumstances" to which Grandey alludes. As the U.S. Nuclear Regulatory Commission's deputy director for engineering Jack Grobe put it, "This is beginning to feel like an emergency drill where everything goes wrong and you can't, you know, you can't imagine how these things, all of them, can go wrong."[88] The earthquake and tsunami caused both flooding and station blackout, ultimately leading to meltdowns at three of the four reactors; winds and currents carried radioactive materials far beyond the plant itself. By the summer of 2011, the government announced that "up to 1,500 square miles (four thousand square kilometers) had been contaminated to the extent that it may require cleanup."[89] And as the authors of *Fukushima: The Story of a Nuclear Disaster* report, "Radioactive substances from Fukushima Daiichi ultimately would be detected in all of Japan's prefectures, including Okinawa, about one thousand miles from the plant."[90] Where to put all of the contaminated topsoil and vegetation that must be removed in this "cleanup" is not clear.

And despite the passing of years, the Fukushima site itself continues to be a problem. The "shut down" process is presently projected to take at minimum thirty to forty years.[91] Among the present challenges is water. With the reactors desperately in need of cooling, TEPCO used seawater, which further damaged the units. Subsequently, groundwater flowed into the site, mixing with contaminated materials such that the water has now itself become a hazard, contained in ever-filling vessels on site at the reactors. As of 2015, TEPCO is at work on an "ice wall" that is to contain the contaminated water in situ until 2020, ironically using, according to the *Guardian*, "enough electricity every year to power 13,000 households."[92] To date, this technology has failed to freeze the irradiated groundwater; in July 2014, Fukushima manager Akira Ono optimistically predicted that it would be successful by March 2015, but as of mid-March, the wall is still in progress. And not all participants are confident that an ice wall is the right response. The *Guardian* cites Dale Klein, former head of the U.S. Nuclear Regulatory Commission and a senior adviser to TEPCO, who is reportedly "not convinced the freeze wall is the best option." "What I'm concerned about," he clarifies, "is unintended consequences. Where does that water go and what are the consequences of that?"[93]

Sustaining investment and interest in the ravages of slow violence, the aftermaths of Fukushima, those unseen effects that persist alongside the various new crises of the day, requires, even for those who are direct victims of it, an act of creative imagination. Most toxins—including radiation, but also other chemical pollutants, from dioxin to lead to PCPs—are not visible without specialized equipment and expertise. The nuclear industry, clearly, must bet on this invisibility, hoping that, as the crisis at Fukushima fades from the headlines (if not from the contaminated groundwater), demand will rebound, both for reactors and for the uranium that fuels them. At present, uranium seems still to be pegged to the fate of Japan, as prices rose optimistically in early 2014 on the release of the Japanese government's draft energy plan that included getting reactors back on line, and sank, subsequently, as the reality of new regulations seemed to slow the process. As mining analyst David Sadowski reports, following the insecurities in the market, "The uranium companies that have been hit the hardest by the weak price environment are those with projects in the pipeline—development-stage projects"[94]—which, presumably, would include Alberta Star. Certainly, in any case, were the site at Port Radium to become once again a productive mine, it would not be manned by miners armed with picks, nor would the ore be carried in leaking burlap sacks; no doubt there would be environmental risk assessment and monitoring of the radiation levels in the water and fish. Nevertheless, one wonders what the future will say of the systems of risk management that govern the higher-tech mines of today. In the Fukushima accident, risk managers had apparently ensured that the plant would withstand either an earthquake or a tsunami, but not both, even though the latter often follows as a result of the former.[95] As spent fuel rods sit in cooling tanks on site at reactors worldwide, we might wonder whether earthquakes and tsunamis are the only ways the systems might break down.[96]

PORTING THE PARADIGM

The new fate of Port Radium may seem, at least for the moment, to render mining (though not, of course, its toxic legacies, which, government assurances notwithstanding, persist) a thing of the past, a minor moment in Canadian history. But seeing Port Radium as a paradigm means porting this example beyond the particularities of time and place. Important as it is to document what happened, stories captured powerfully in the medium of documentary film, strictly historical accounts risk isolating the event. It is perhaps here that experimental drama of the sort staged in *Burning Vision* offers something different, a way to carry the paradigm forward, loosening the link between representation and the real, for though the play's subject matter is nuclear, Clements's abstract, symbolic, and suggestive treatment also renders the play more mobile, as, restaged in new contexts, it might speak to new risks, taking it beyond the more journalistic reporting of a conventional documentary film. As Catriona Sandilands notes, literature is useful for environmental staging in part through its very nonreferentiality: "Environmental literature may indeed be a creative act of enduring worldliness for public discussion and action, but its very qualities of literariness remind us that, in developing practices of representative thinking around it to form and evaluate opinions, this world is still not nature itself."[97] Thus, though it is useful to key *Burning Vision*'s more symbolic and suggestive images to the more literal history presented in Blow's *Village of Widows* or Henningson's *Somba Ke*, the play also offers a vehicle for considering other sorts of hazards, including the ravages of the oil and gas industries into which Alberta Star seems to have moved.

In "Atomies of Desire: Directing *Burning Vision* in Northern Alberta," for example, Annie Smith describes her own experience in directing the play as activist intervention:

> A debate had begun in 2008 about building nuclear power plants to provide power for the tar stands in the communities of the Peace River region, both in British Columbia and Alberta. I felt that *Burning Vision* might bring some historical perspective to this debate, bringing to light the devastation unleashed by the willful ignorance and greed of a previous era, and reminding us that these same attributes could prove equally harmful today.[98]

The issues of environmental concern here are not strictly the same as those represented in the play—nuclear power plants are not nuclear bombs; tar sands are not tailings—but by staging the play in these new circumstances, Smith asks her audience to read analogically, to see in these new contexts a repetition of a structure from the past.

Smith casts this repetition in a moral register ("willful ignorance and greed"), suggesting, in her "director's notes" in the program, that the audience's "ignorance" was in fact even more "willful" than that of the characters in the play:

We do know that the people at Fort Chipewyan are suffering unprecedented rates of cancer. We do know that the wildfowl and other wildlife are having their habitat destroyed due to the oil sands extractions. We do know that the oil sands extractions are using and polluting huge amounts of water, which is, in turn, poisoning vast tracts of land and our water systems. We can't really say, like Gilbert LaBine, "I didn't know."[99]

Smith's assertion of knowledge—"we do know"—is, however, as much political as it may be factual. We may know, but so did Gilbert LaBine. We also, and also like him, don't know. As laypeople, we can certainly intuit, we can make observations, we can draw logical conclusions, but grappling with the complexities of risk society also means not knowing but being responsible anyway. In this, too, we are like Clements's Bros. LaBine.[100]

Setting *Burning Vision* to work in a new political and historical context, Smith in effect abstracts the nuclear content, rendering the play a mobile metaphor or allegory for risk more generally. And she closes by drawing an analogy specific to the particular form of "staging" that drama offers, suggesting that the irreparable losses of environmental hazard in time replicate the intangible "ecology" of live theater performance:

> There is no way to call back the creative energies of the performance, just as there is no way to call back the destruction caused by the bombs that were dropped on Hiroshima and Nagasaki, or the thousands of lives that were lost. And so, too, there is no way to call back the land, the water, or the creatures that have been destroyed by the tar sands developments of our time. But theatre can bear witness, confront us with our ignorance, and—if we pay heed—guide our actions to preserve life, not destroy it.[101]

As her last comment suggests, even if each individual live performance is fleeting, the existence of the play suggests its iterability; performances can be restaged, as she has done, in new contexts, generating new political energies capable of inspiring new forms of critical and political engagement. And here, Déline might offer yet another paradigm, for if, in their visit to Hiroshima, Déline Uranium Committee members offer a model for taking responsibility even when there is no foreknowledge of the consequences of one's actions, a more recent document might offer another model, this time in a more precautionary vein. If, in the context of Canadian tar sands oil extraction, the ethos seems to be "We'll go until it becomes apparent that we've gone too far, and then we'll stop. . . . The canary has to die," community members in Déline have crafted an alternative vision in a 2005 document called "The Water Heart," which asserts, according to Chris Wood, that "any development must preserve the environment—and particularly Great Bear Lake's pristine water—substantially as it was before the first soil was turned for the project, and not merely limit its damage to a southern idea of what is reasonable."[102] Such preservation would need, of course, to attend to the rhetorical cleansing that inevitably accompanies such development, lest the quibble of "substantially" open the floodgates to new contaminations. "Canaries," after all, die daily in the tar sands, where "too far" is clearly a boundary already surpassed.

And if the history of Port Radium might be a paradigm of environmental threat—defined by van Wyck as "Something happens here and now because it really happened over there and then"—this only becomes more apparent in an age of climate change. Like those islands in the Pacific, first exposed to fallout in nuclear testing and now affected by rising sea levels, the people in the north are also disproportionately affected by climate change's uneven effects. Canada's north, as Lorien Nesbitt reports, "is experiencing some of the most extreme climate changes on Earth. Annual temperatures have increased by 1–2 degrees C over the past 50 years—twice the global average."[103] Nesbitt goes on to chronicle the changes that have occurred and are predicted to occur with greater frequency and intensity as temperatures continue to rise, from the melting of permafrost to the increasing turbidity of the water to the greater precipitation (but also greater evaporation) to the wind speeds and even length of the day, as increasing cloud cover changes the angle of light in the sky. These changes are happening in Déline because they really happened "over there and then," in the past burning of fossil fuels that has transformed the environment, including the very basics of life in the north, the fish in the lake, the caribou on the land, now swimming in turbid waters or fighting the heavy snowfalls.

SPEAKING TOO LATE AND TOO SOON

As Clements's *Burning Vision* comes to a close, the final words of the Seer's prophecy warn us of a future even yet to come: "This burning vision is not for us now . . . it will come a long time in the future. It will come burning inside."[104] Complicating the logic of fast violence—the spectacle of the flash of the bomb or the crisis of the meltdown—the prophecy emphasizes the slower violence, the "burning inside" of mutation or carcinogenesis or toxification or climate transformation. And, in the face of an "even yet to come," the play also opens questions for us now, inhabiting the ambiguous future to which the See-er refers. The discourse of risk management is in effect an attempt to control the future, to domesticate what is to come by extrapolating what has come before, but in risk society, this speculative imaginary no longer seems to provide reassurance, as the very grounds of our pasts and presents—what we thought we knew—seem uncannily to change before our eyes. As Nick Mansfield puts it in the context of climate change, "We will make our future in relation to the limits we have ourselves made, the past we have engineered not intentionally but unwittingly, the past now coming back to haunt us from the future."[105] Though "the past" here is something like CO_2 emissions, Mansfield's description resonates with the nuclear issue as well, as the industry promises its messianic return. As Mansfield reminds us, invoking Derrida's *Specters of Marx*, the specter that haunts can be for good or ill; in the case of nuclear, though, the specter might better be described in terms of zombies than of ghosts. As Beck suggests: "The nineteenth-century concept of risk, when applied to nuclear energy, is a zombie concept, a category of the living dead that blinds us to the reality in which we live."[106] When the paradigm of environmental risk is that "something happens here and now because it really happened over there and then"—with the understanding that all terms in play, from "happens" to "really" to "because," are all open to question—risk management, that undead actuarial imaginary whereby threats could be anticipated and insured against, is clearly inadequate.

Revelations are ongoing in risk society, and any narrative of environmental risk or catastrophe inevitably ends too soon. Describing the experiences of the people in Déline, Peter van Wyck notes the uncanny way in which recognition of the slow violence of uranium transformed the landscape: "Dust and dirt became 'tailings.' Hands not washed became precautions not taken. Caribou and fish were freighted with risk. . . . In this way, the past, their past, was itself rendered toxic by virtue of a retroactive catastrophe of knowledge."[107] Or, as Beck puts it, "New knowledge can transform normality into threat overnight."[108] Indeed, we might take the subsequent history in Déline, the report's happy rewriting of hazard, the good news that, as one of Henningson's informants tells us, "the environment was safe and clean. The fish were safe to eat, the water was safe to drink in the Port Radium area," as paradigmatic of the ways in which revelations can reverse course, inverting Beck's formulation, as new regulation can transform risk into normality overnight. Such a move also characterized the aftermath in Fukushima, for, in April 2011, the government raised the acceptable level of exposure to radiation for children from 1 to 20 millisieverts per year.[109] With regulatory fiat, toxic playgrounds thus became safe again—in this case effecting a rewriting of the present and future, for whatever dangers there might have been for those children (given the slow operation of radiation and the complex etiology of cancer) the changed regulations were to guarantee that the accident would no longer have been the cause. Here, that intangible, invisible nuclear risk is "staged" by industry, by regulators for laypeople, the same landscape rendered toxic or safe by those experts in whose hands our pasts, presents, and futures seem to be placed.

But here too we might see in Clements's Bros. Labine 1 an object lesson, for while he questions the knowledge of experts as an alibi for abdicating responsibility, we might read expert knowledge in a more precautionary register. Risk management is, in effect, based in analogy: past events will have been like future ones, and thus risk can be managed, mitigated, and insured against. What escapes the analogy—those ways in which the future is not like the past, or in which the past itself is changed in the face of present revelations—is not factored into the calculus of responsibility. The precautionary thinking associated with risk society, however, opens up those questions of analogy and responsibility, as diverse technologies end up being analogous, not in their predictable risks, but in their unprecedented unpredictability, a moment in which the gaps in the analogies, their clearly political nature, become increasingly open to public view. In the case of the Japanese government's shift from 1 to 20 millisieverts, for example, the analogy was the child to the adult nuclear worker, for whom the latter regulation had originally been made. But Fukushima parents offered their own intervention when they gathered a bag of contaminated playground topsoil and delivered it to the upper house of parliament in protest, political theater, captured by the international media, that opened again to public debate questions of precaution and risk.[110] Following public outcry, by the end of May 2011, the government had reverted to the original standard of 1 millisievert per year, and Tokyo committed to help fund the costly removal of contaminated soil from schoolyards.[111]

Returning to Derrida, Nick Mansfield reminds us that the "to come" of the incalculable future need not be positive, the salutary promises of democracy or justice generally associated with Derrida's work. "There is," he argues, "no reason to believe that what open-ness is open on is good."[112] As Ulrich Beck warned in a pre-Fukushima piece in the *Guardian*, the message of that moment in the Anthropocene seemed to be "all aboard the nuclear power superjet. Just don't ask about the landing strip." The landing strip to which he was referring was the still open question of nuclear waste storage, and he alludes there to the questions raised by the Waste Isolation Pilot Plant (WIPP), a deep geological repository for some of the most persistent of nuclear wastes, in which even the seemingly nontechnical matter of signage became impossibly complex, a subject treated at length in Peter van Wyck's marvelous *Signs of Danger* (2004).[113] Since whatever marker is used at the site must not only endure but signify for some ten thousand years into the future, the problems of marking are fairly intractable. But as van Wyck's book reminds us, thinking only of that long term ought not to divert our attention from the ongoing catastrophe that the site represents in the present. Vaults leak, as workers at the WIPP learned in February 2014 when the alarms attached to air monitors at the site were set off by excess radiation escaping to the surface.[114] And, in the wake of Fukushima, we were reminded, again, that the danger in nuclear power is not just the "landing strip" but the "superjet" itself. As Beck put it after the accident in 2011, in an uncharacteristically confident prediction of the incalculable future, "If anything is clear, it is that another nuclear disaster is a certainty."[115]

For the risk critic, thinking in the precautionary registers of François Ewald or Ulrich Beck, it might be as important, though, to recall the opposite. The incalculable in Beck's work is generally cast in the register of uninsurability, of total runaway catastrophe and colossal disaster—meltdowns, whether of nuclear power plants or polar ice packs. But critique of risk can speak too soon in its pessimism as well as in its optimism. The narrative with which I began this chapter, Terry Tempest Williams's "The Clan of One-Breasted Women," was originally published in January 1990. In it, she chronicles the failures of the U.S. government to acknowledge responsibility or offer restitution for the exposures to radiation posed by mining of uranium or above-ground atomic testing. While she notes a brief victory in a case in 1984, she concludes with the Supreme Court upholding a lower court's decision based on a doctrine of sovereign immunity: "To our court system it does not matter whether the United States government was irresponsible, whether it lied to its citizens, or even that citizens died from the fallout of nuclear testing."[116] Such seems to be the final word on the matter, delivered by the Supreme Court, absolving the government of any liability for the cancers suffered by miners, ore carriers, or downwinders. As it happened, however, a mere nine months later, the United States passed the Radiation Exposure Compensation Act, legislation that, while inadequate in many ways, and no doubt passed on the assumption that it would forestall further costs of litigation and claims, did mean that the government accepted responsibility and, to a limited extent, compensated victims.

Of course, the Radiation Exposure Compensation Act is retrospective. It does not cover losses for those, say, presently living near old mine sites, many of which are still toxic, or losses to come, as the nuclear industry recovers from Fukushima, but the act has been amended several times, each iteration taking on a wider responsibility. If the Cold War's collateral damage has been thereby acknowledged, in however limited a fashion, one wonders what the future compensation might be for those victims of climate change or "the war on global warming," those who suffer directly from the rising sea levels, droughts, or megastorms, and those who fall prey to the side effects of trying to mitigate those primary hazards, the once again newly acceptable risks of future Fukushimas and Port Radiums or the unknowable risks to come represented by geoengineering. Liability, in these new contexts, is unlikely to be traceable to particular governments or corporations. The very "global" nature of climate change means that the phenomenon of organized irresponsibility has truly become total.

Precautionary thinking of the sort that "being sorry before you have to say you are sorry" represents is about acknowledging the incalculable; it is an expansion of the concept of responsibility. Such precaution resonates with Derrida's notion of futurity. To do justice to the incalculable future, one must be ready to be responsible for the consequences, unpredictable and radically other as they may be. For Derrida,

> Incalculable justice *requires* us to calculate. . . . Not only *must* we calculate, negotiate the relation between the calculable and the incalculable, and negotiate without the sort of rule that wouldn't have to be reinvented there where we are cast, there where we find ourselves; but we *must* take it as far as possible, beyond the place we find ourselves and beyond the already identifiable zones of morality or politics or law, beyond the distinction between national and international, public or private, and so on.[117]

As environmental issues perpetually remind us, "the place where we find ourselves" is nearly always already "beyond" those distinctions. In this context, we might mobilize, as van Wyck at least implicitly does, Port Radium not only as a paradigm of environmental disaster, but also as a paradigm of the kind of ethical engagement necessary for taking responsibility for environmental harm. This kind of calculation certainly can be one role for literature: the telling and retelling of narratives of risk in the hopes that the future may be written differently. Terry Tempest Williams concludes her book with a description of a protest at the Nevada Test Site where she is detained and frisked by the police. When they discover pens and paper, she affirms: "Weapons."[118] And whether or not climate change is inherently more "cinematic," as Tom Cohen suggests it is, both in its modes of representation and in its very materiality ("no oil, no hyperindustrial technoculture, no photography as we know it, no cinema"),[119] films, too, clearly have a role to play in the staging of risk, itself an ironic twist in the narrative of uranium, for, as Peter van Wyck reports, "It is said that prospectors would use photographic film to find pitchblende by its trace."[120]

AFTERWORD. Writing "The Bomb": Inheritances in the Anthropocene

> *The concept of responsibility has no sense at all outside of an experience of inheritance.*
> —Jacques Derrida[1]

The Doomsday Clock of the *Bulletin of the Atomic Scientists* arrives in the second nuclear age as an inherited artifact, a legacy and a kind of time machine, for like the trinity of scientists imagined in Lydia Millet's *Oh Pure and Radiant Heart*, the Clock represents a moment when the fears and hopes of the nuclear were largely to come. The June 1947 issue in which the Clock first appeared is full of the seemingly real possibilities of world government and with it a world without war, the belief that, still, the bomb might have the potential to usher in an era of peace. Clearly, we have inherited that era's weapons. And we have inherited the results of the detonations of those weapons, the traces of which have recently led Paul Crutzen, coiner of the term "Anthropocene," to recast its origin as the origin of the atomic age.[2] And, even if the hopes of the scientists' movement, of global cooperation and abolishing war, seem as anachronistic as Millet's Fermi, Oppenheimer, and Szilard, hope itself surely remains, as does that expression of the scientists' conscience, the *Bulletin,* still in print some seventy years after Hiroshima and Nagasaki revealed their gadget to the world.

But we have inherited many other things as well: the petrochemical industrial complex (including plastics production) that also exploded in and after the war; the resultant new climate that seems to change before our very eyes; the biotech, nanotech, and infotech revolutions that promise to transform our world, for better or for worse; the willingness to risk environmental catastrophe for dubious gains, as in the recent instance of fracking, with its groundwater contamination and temblors. And we have inherited the systems of accumulation and inequality that are imbedded, inevitably, in and around those technologies. Alluding to this expanded inheritance, the *Bulletin*'s January–February 2007 issue, with its new "second nuclear age," attempts, in effect, to resignify the Clock, the same graphic now meant to capture less explosive, in some cases slower, but no less deadly apocalyptic risks, but the inherited futurity implied in the Clock's midnight belies the ways in which these are disasters at once in motion and to come, as "climate change and advances in the life sciences" are clearly upon us—in much the way that the nuclear in the first nuclear age was present, in testing and accidents, even when it was figured as "fabulous" and futural. What may have made sense as a warning image for the atomic scientists in the 1940s needs now to accommodate the various speeds of violence, the varying degrees of risk in what may indeed be "one [uneven] world," but is no closer to world government, and the past, present, and future of the becoming-catastrophe of global risk.

As though cognizant that the single icon might not be able to bear such weight, the editors of the 2007 issue of the *Bulletin* included not only a number of articles that were to clarify this new era of risk but also an illustration by John Hendrix that spelled—or rather drew—out the extent of these threats in an imagined apocalyptic landscape.[3] Here, a large eagle sits at the center of a roaring river, holding a nuclear missile, an ominous cloud with the word "Doomsday" emitting from its mouth, and a precipitous waterfall ominously threatening. Also in the water—and headed for the brink—are other threats, represented in words and pictures, "radiation," "fall out," "extinction," "famine," "global warming," "flooding," "plague," "war," "genetics," "DNA," "mutation." Though there are some UN workers on the bank of the river, a section called "Prevention," they are clearly distracted and overwhelmed, trying to force an octopus of problems back into a chest or single-handedly holding up a dam that, in any case, is only blocking a tributary to the main stream. Missiles, meanwhile, fall from the sky, and a mushroom cloud rises on the far, dark horizon.

Hendrix's illustration is arresting, and as a symbolic representation of the second nuclear age, it goes some way toward correcting the futurity of the Doomsday Clock. The disaster here is both present and to come: GMOs out of the "Little Shop of Horrors" snap at scientists who have strapped a human subject under a giant microscope; penguins stand in disbelief on a single remaining piece of sea ice; cities are swept downstream; and altogether they are headed for a precipitous future. Hendrix's drawing aptly illustrates the moment it was commissioned to represent, but the second nuclear age is characterized by its own blind spots. The very sci-fi surrealism of the scene might foreclose an analysis of the ways in which this world is not wholly new, the ways in which the first nuclear age was also charac-

6. John Hendrix, "Doomsday." (From *Bulletin of the Atomic Scientists* [January–February 2007]. Courtesy of John Hendrix.)

terized by catastrophe. I am struck, too, though, by the excessiveness of representation in this issue of the *Bulletin*. There is something in the iterative and collaborative quality of the Clock, the text, and the illustration that suggests the difficulties of staging a world at risk—and the necessity for multiple voices, media, and forms, all marshaled for a more precautionary politics.

Inspired by the *Bulletin*'s marriage of text and image, I would like to close with a somewhat different example of visual art: the collaborative murals of Japanese artists Iri and Toshi Maruki. Their work was, like the Doomsday Clock, inspired by the first atomic bomb, and it also, anticipating the "second nuclear age," saw in that weapon a pattern or paradigm that was repeated in further disasters. But whereas the record of the Clock's shifting hands offers us a history of the past's future, the Marukis' murals provide an archive of the future's pasts. Staging—that practice of representation—of risk is generally oriented toward the future, offering a way to grasp its "becoming-real," but the practice of risk criticism must be as attentive to the ways in which past risks have played out, with their unforeseen wagers and incalculable odds. The Marukis' artistic practice provides an archive of the past that seems to imagine the perspective of posterity, chronicling our inheritance while imaging a future capable of charting a different course.

The Marukis' work is not in itself futural; they do not speculate on coming catastrophes. Their murals were always commemorative, oriented toward a past that they felt must be represented. Taken as a whole, though, their project offers a model of an iterative practice of returning to "the bomb" with a difference, drawing analogical connections among diverse horrors and persistently returning to their own work, to add captions, give interviews, or appear in documentary film, producing what Charlotte Eubanks has recently described as a kind of performance art, a project itself subsequently extended by other artists and writers.[4] As such, their work presents a model, both for the staging of catastrophe and for the staging of the staging of catastrophe, the latter a reflexive practice that might inform a risk criticism as well. The Marukis' work is at once collaborative and invites collaboration. As Iri put it, "Our joint works are not my paintings or Toshi's: they are ours. And not ours alone, because they are created with the help of many people."[5] In that spirit of collaboration, of inheritance, *Risk Criticism* aims to participate in and carry forward elements of that political project.

HOW I LEARNED TO START WORRYING AND LOVE LITERATURE

Toshi Maruki's art was in fact central to the development of my own senses of hope and fear, for my first real encounter with the history of the bomb was her haunting picture book *Hiroshima No Pika* (1980). This and Eleanor Coerr's 1977 *Sadako and the Thousand Paper Cranes*—these were the stories that introduced us, children of the late Cold War, to the horrors of the bomb, a bomb that loomed large in my imagination, as I did walkathons and bikeathons, distributed literature and asked for donations for the nuclear freeze campaign for which my parents worked, the modest goal of which was to "freeze" the production of nuclear weapons. It was the age of overkill, of full-blown MADness: No duck-and-cover drills for us; the bomb was not deterred by desks; we weren't taught to protect our delicate eyes from the flash. Toshi Maruki's *Hiroshima No Pika* depicted for her audience the horrors of that first atomic bomb. The illustrations were graphic; the story terrifying. A young girl, and one not unlike myself at the time—one minute eating breakfast with her family; the next minute in chaos. I imagine my parents took to heart what Toshi said in her "about this book" note, that "it is difficult to tell young people about something very bad that happened, in the hope that their knowing will help keep it from happening again."⁶ Hopelessly romantic, of course, but also no doubt one reason why I went on to believe, myself, in the power of literature and art.

The story of *Hiroshima No Pika* is devastating in its brutal sparseness, but it is the art that truly lingers in and haunts the imagination. And this is not surprising, given that Toshi Maruki was first and foremost a visual artist. Her children's story was inspired, at least in part, by a Hiroshima survivor (or Hibakusha) who shared her own story with Toshi after seeing the murals for which Toshi and her husband Iri Maruki were famous, a lifelong project that, too, was inspired by the hope that knowing about very bad things that happened in the past would prevent them in the future. In the immediate aftermath of the bombing in Hiroshima, Iri and Toshi Maruki had journeyed to the site; had seen the horrible suffering and destruction; had endured, with the survivors, the black rain that subsequently fell. After this experience, as the events remained shrouded in official silences, the artists, who had previously established independent careers, Toshi in oils, Iri in ink and watercolors, collaborated on the first of what would become the Hiroshima panels, immense murals depicting the human suffering so often occluded in official accounts of property damage or number of casualties.

Their collaborative project, which lasted over forty years, produced fifteen Hiroshima panels, each titled thematically according to the subjects treated, as well as murals depicting other atrocities, the radiation poisoning of the Japanese fisherman aboard the *Lucky Dragon* who had the misfortune to be in the path of the fallout from the Bravo test in the South Pacific, the deaths of American POWs and Koreans in Hiroshima, and ultimately the Rape of Nanking, Auschwitz, and even Minamata disease, the mercury poisoning caused by industrial wastes that contaminated the sea life and those who consumed it in Minamata, Japan, in the 1950s and 1960s. The project culminated with a mural simply called *Hell*, into which they painted, not just Hitler and Truman, but also themselves, because, as Toshi explained, though they had not dropped a bomb or killed anyone, they had not been able to prevent war.⁷

Each of these murals is different, of course, but the overall project is clearly unified, with the atomic bomb as a centerpiece and the subsequent horrors as variations on that theme. Different as Hiroshima and, say, Minamata might be—the former a wartime atrocity intended to kill, the latter a peacetime act of reckless indifference and callous negligence—the continuities in style across the murals suggest that there is a basic similarity, even analogy, between them. The impetus for painting Hiroshima was personal: they had observed the aftermath, and they felt a sense of responsibility for that witnessing; the other atrocities, though, were inherited secondhand, as observers of their murals challenged them to consider "war" in a much broader context, "from the frenzied killing of others in war to the slow killing of the environment," as John W. Dower describes their work.[8] Or, as Iri Maruki put it, in an expression of the sense of responsibility that animates their work, "Problems presented themselves and we could not turn away."[9]

Their project is thus shaped by expansive vision, both of shared catastrophe and of resultant responsibility, commitments of the sort that might animate the work of risk criticism as well. And in the service of the larger ethico-political imperatives of their work, the Marukis themselves also took risks, political, ethical, and artistic. They certainly risked their own reputations by treating contentious themes, by depicting what was officially occluded, and by refusing to align with any particular side. They risked seeming to co-opt other catastrophes that weren't strictly their "own." They risked, in painting atrocity, reproducing the very violence they opposed, or producing the kind of apathy that can follow from a world on the verge of apocalypse. And the art itself was a risk, both in its collaborative production and in its display, as the artists persistently crossed genre and medium, art and politics, all of which, they insisted, were risks worth running in the service of the larger purpose of their work, not, Iri insisted, "to edify," but to "paint that reality," the "hell" in which "all living human beings" reside.[10]

THE ART OF STAGING CATASTROPHE / STAGING THE ART OF CATASTROPHE

The Marukis' murals themselves, clearly, stage catastrophe, and as such are a means of reflecting upon atrocities of the past and their legacies today. Their "Hell" offers a potential iteration of Beck's reflexive second modernity, a world in which we create the environment that now seems to confront us, with its ozone holes and climate catastrophes, toxic clouds, radioactive soil, endocrine disruptors, superbugs and superweeds. Confronting a hellish reality, the Marukis offer something like the reflection that might follow from this reflex, and they have, in turn, inspired further reflection, as those witnessing their work have gone on to represent it again, in new iterations and new forms, a recursive movement that is, implicitly, authorized by their original project. Toshi's children's book, of course, blends text and image, seeming to confirm William Greenway's contention that "today we seem to find art and literature collaborating only, though wonderfully, in children's literature" (47). But this collaboration also characterizes the murals, for among the risks run in their work is the blurring of artistic media. Though Iri reportedly initially objected, Toshi wrote captions for their panels, each of which is to explicate and further the project of the painting. Here, for example, is the caption for *Atomic Desert*:

> There was no food, nor medicine.
> Houses were all burned,
> the rain came in.
> No electricity, no newspaper to read, no radio,
> No doctor.
> Both the dead and wounded
> were food for maggots,
> and swarms of flies buzzed.
> The odor of the corpses was on the wind.
> The Atomic Bomb exploded in human hearts
> as well as upon human bodies.
> Heedless of naked and ragged skin,
> they would search for lost children
> day after day.
> Even now,
> human bones are found in the soil
> in Hiroshima.[11]

Somewhere between caption and ekphrastic poem, the words describe something at once in and adjacent to the scene, adding sounds and smells, noting what is not depicted (the electricity, the newspaper, those marks of everyday ordinariness that are so starkly absent in the picture) as well as describing what is. Bringing us into the present—"even now"—the caption calls us to see these images in the seemingly cleansed soil of the present, where the materialities of past atrocities remain, even if they are not clearly seen.

Toshi reported that she received criticism for this practice, her seeming "contamination" of their art with the textual, but she insisted, on ethico-political grounds, on the compatibility of these media:

> To be sure, I ran into the criticism that I infringed on the purity of the paintings. That may be so. But what wasn't said completely in pictures I had to communicate orally. And what I couldn't say completely orally, I had to write. By hook or by crook, I wished to com-

7. Iri and Toshi Maruki, *Atomic Desert*. (Courtesy of the Maruki Gallery for the Hiroshima Panels.)

municate the truth of Hiroshima to as many people as possible. How can that be impure? What after all is art?[12]

Toshi's last question ("What after all is art?") seems, in context, fairly rhetorical. For the Marukis, art is an ethico-political project of remembrance that contains a promise for the future; it is an omnivorous and flexible practice that draws, pragmatically, on the tools at hand. To take the Marukis' work seriously is, then, as Charlotte Eubanks has suggested, to read the murals as "but one facet of what is primarily a performance-based art form,"[13] with the Marukis' own enacting—orally or in print—of the ethical principles on which their work is based as a component of the larger project.

If an essential part of the project of the Hiroshima murals is the Marukis themselves, what, Eubanks asks, will become of the project now that both artists are deceased (Iri in 1995 and Toshi in 2000)? "For decades," she argues, "the Marukis played the dual roles of preacher (giving vocal interpretation) and contrite sinner (exposing their own wrongdoing), thus providing a model for others' engagement. With their deaths," she warns, "part of the artwork is in danger of being lost." Thus, she continues:

> The questions now become: How does the artwork change if its meaning is reduced to the positive content of the murals, without including the performance work of . . . interpretation? With the primary actors gone, what techniques are available for ensuring the continued, if altered, performance of the work? And who might step in to take over the work of vocal interpretation?[14]

Clearly, Eubanks's own essay, published in *PMLA*, goes some way toward resituating their project in their performances, and she acknowledges that "art critics and academics have their place"; she notes too that "knowledge is also enacted by artists, activists, educators, and students," work that is encouraged in part by the institution of the Maruki Gallery in Japan.[15] These multiple and diverse examples suggest that we all might inhabit new roles in a larger performance piece—and one that might include the project of risk criticism that I have hazarded here. As a bridge between their project and mine, I turn, in conclusion, to an artist who offers one possible mode of engagement, American poet Ronald Wallace, who, in two poems, "The Hell Mural: Panel I" and "The Hell Mural: Panel II," both published, originally, in *Prairie Schooner* in 1988 and subsequently in his collection *The Makings of Happiness* (1991), engages and furthers the project of "painting the bomb."

As a kind of metarepresentation of the Marukis' murals, Wallace's poems offer an occasion for thinking about staging and restaging, both of past catastrophe and future risk, and thus offer something like a poetic rendering of risk criticism, both in their subject matter and in their form. As verbal representations of visual art, Wallace's poems, like Toshi's captions, draw on the tradition of ekphrasis. But if they are examples of this genre or mode (depending on who is defining it, ekphrasis appears to be either or both), they also challenge it. The murals, as Eubanks's analysis suggests, now stand alone, without their accompanying "performance work of . . . interpretation," and, on display in a gallery, cast as vehicles for memory, they risk a certain stasis, especially given the kind of fetishizing of visual art that Toshi so adamantly resisted. Ekphrastic poetry, if we take Murray Krieger's 1967 essay "Ekphrasis and the Still Movement of Poetry; or, Laokoon Revisited" as a touchstone, is precisely about stillness, about timelessness, the arresting of the forward drive of narrative in a kind of spatial formalism. Of course, subsequent critics have troubled this definition. As James Heffernan notes, "Krieger's theory of ekphrasis would hermetically seal literature within the well-wrought urn of pure, self-enclosed spatiality, where the ashes of new criticism now repose."[16] Countering both Krieger's notion of ekphrasis as a freezing of time and space and Wendy Steiner's reading of the ekphrastic poem as a "pregnant moment," Heffernan argues that "ekphrastic literature typically delivers from the pregnant moment of graphic art its embryonically narrative impulse, and thus makes explicit the story that graphic art tells only by implication."[17] "The story" here is presumably the story implied in the art, not the story of the art's production, but if the Maruki murals are, as Eubanks convincingly suggests, "one component of a performance based art form," "roughly analogous," she argues, "to the way that set design is one aspect of theater,"[18] then the "story" that needs to be told must take in the broader theatrical context as well as the murals themselves.

A glance at Wallace's "Hell Mural: Panel I" suggests the extent to which these poems highlight this more narrative element, for what is described in the poem is, not just the action that is embryonic in the still image, but the painting and unveiling of those images:

Iri and Toshi Maruki are "painting the bomb."
Their painting, they say, will comfort the souls of the dead.
"It's a dreadful cruel scene of great beauty,"
Toshi says. "The face may be deformed but there's kindness
in a finger or a breast, even in hell."
The Hell Mural spreads over the floor.

Iri stretches naked on the floor,
painting. He remembers Hiroshima after the bomb —
the bodies stacked up, arms outstretched toward hell,
nothing he could see that was not dead,
nothing that cared at all for human kindness,
nothing that wept at such terror, such beauty.

> Now a brush stroke here, a thick wash there, and beauty
> writhes and stretches from the canvas floor.
> He wants his art to "collaborate with kindness,"
> he wants his art to "uncover the bomb."
> But no lifetime's enough to paint all the dead
> or put all those who belong there in hell.
>
> "Hitler and Truman," he says, "of course are in hell.
> But even those of us who live for beauty
> are in hell no less so than the dead."
> (He paints himself and Toshi on the floor.)
> "All of us who cannot stop the bomb
> are now in hell. It's no kindness
>
> to say different. It's no kindess
> to insist on heaven; there's only hell."
> Toshi adds bees and maggots to the bomb,
> and birds, cats, her pregnant niece, the beauty
> of severed breast and torn limb on the killing floor.
> "In Hiroshima," she says, "we crossed a river on the dead
>
> bodies stacked up like a bridge. Now the dead
> souls must be comforted with kindness.
> Come, walk in your socks across our floor,
> walk on the canvas. (A little dirt in hell
> almost improves it.) Can you see the beauty
> of this torso, that ear lobe, this hip bone of the bomb?"
>
> Iri and Toshi Maruki, in "Hell," are painting the bomb,
> the mural on their floor alive with the thriving dead.
> Come walk on their kindness, walk on their troublesome beauty.[19]

The moment here is far from still, as the artists speak—in the quotations—stretch on the floor, painting, and even invite us, the reader/viewers to walk into the world of the painting ourselves, a mural so in process that it can accommodate new material seemingly indefinitely. But if we take Heffernan's general principle that most broadly, "ekphrasis uses one medium of representation to represent another,"[20] Wallace's poem adds yet another layer, for though the poem does not mention this explicitly, the scene that it describes is not exactly one imagined whole cloth by the poet; rather,

8. Photograph by Motohashi Seiichi. (Courtesy of Maruki Film Project.)

"The Hell Mural: Panel I" in all of its quasi-documentary mode, was in fact inspired by a documentary, John Junkerman's 1986 film *Hellfire: A Journey from Hiroshima*, which framed a retrospective on the Marukis' work by documenting their production of their *Hell* mural.[21] Wallace's poem is thus a representation (poem) of a representation (film) of a representation (mural plus performance).

Describing the origins of "The Hell Mural: Panel I," Wallace explains, "The poem had its genesis in a short documentary film I saw about Iri and Toshi Maruki. . . . Their statements about art were very moving, including, among other things, the idea that they wanted to pass their work on to others who would keep it alive in different forms."[22] Viewing the film, one form in which the work is being kept alive, the poet hears an ekphrastic call to recast their work anew. As the filmmakers strive to document the performance art that includes the murals, so too the poem works to capture this project, translating it now from documentary, a genre whose conventions the poem clearly borrows, into a new medium, balancing a desire to retain the original message with an acknowledgment of the levels of mediation through which this message has traveled. Navigating a tricky line between inspiration and co-optation, the poem recasts the often naturalized transparency of the documentary in the highly artificial form of the sestina.

Indeed, insofar as there is anything still in the poem, it may be the sestina form, which, with its arbitrary rules dictating the stanza number and length and repetition of end words, at least gives the impression of persisting rigidly, unaltered by content or historical context. In a presentation of repetition and difference that is analogous to the iterative quality of the Marukis' multiple murals, as the end words (bomb, dead, beauty, kindness, hell, floor) circulate through the lines of the poem, their valence changes, the most mundane of these, "floor"—a reference to the fact that the Marukis painted on the floor—becomes the ground beneath us all, the Hell in which we all must acknowledge we live. The final invitation in the envoi, to "walk on their kindness, walk on their troublesome beauty" draws us at once to enter and contaminate—"A little dirt in hell almost improves it"—to violate those fetishized boundaries of art.

The poet's invitation here is a direct reference to the film, which shows Iri Maruki at the *Hell* mural's unveiling, inviting members of the press to walk in their socks on the mural. In the context of the poem, though, the invitation is broader and can be read as a variation of what W. J. T. Mitchell calls the ambivalence of ekphrastic hope and ekphrastic fear, the former, for Mitchell, the hope that the poem can approximate art; the latter that of generic cross-contamination.[23] In the case of "The Hell Mural: Panel I," the hope seems to be the replication of the ethico-political/aesthetic project in a new form; the fear that such iteration might be appropriation—not just "walking on," but "walking all over," as the phrase goes. The poem offers us the opportunity to enter the mural, the art that has turned the bomb and the dead into something "alive" and "thriving." We cannot refuse the invitation to inhabit the hell that they depict—we are already there—but we must inhabit it well. As we "walk on their troublesome beauty," we surely feel the anxiety of trespass, but also the immense responsibility we have in treading with care and kindness in this tortured terrain.

PASSING IT ON

"The Hell Mural: Panel I" thus commemorates both the staging of catastrophe and the staging of the staging of catastrophe, emphasizing the importance of keeping the memory of the bomb alive, as the bomb itself persists—so many "seeds," as Lydia Millet suggests, "waiting to bloom." But to do full justice to the Marukis' political project is not merely to re-present the act of representation, but also to perform the riskier practice of carrying it forward in the spirit of expansiveness that their original work represents. Indeed, the precautionary spirit in the final lines of "The Hell Mural: Panel I" perhaps prepares us for the risks Wallace takes in "The Hell Mural: Panel II," for here, in a villanelle, the poet not only returns to the not-so-still moment of artistic creation, but extends it, imaging future murals that the artists have not yet painted, but could paint. Many of the themes return in this poem—horror and kindness, the imperative to represent and record—but the iterative quality of the refrains draws our attention to the new project that the poem proposes:

> Iri and Toshi Maruki are painting the bomb.
> Their painting, they say, will comfort the souls of the dead
> in Hiroshima, Nagasaki, Belsen, Dachau, and Vietnam.
>
> Because Hitler and Truman and Nixon with such aplomb
> could order the deaths of millions, then go to bed,
> Iri and Toshi Maruki are painting the bomb.
>
> They draw in bees and maggots, and then go on—
> a nipple here, a finger there, a head—
> to Hiroshima, Nagasaki, Belsen, Dachau, and Vietnam.
>
> Birds, cats, naked men and women spawn
> on the floor in their mural, burning—the beautiful dead
> of Iri and Toshi Maruki's painting, the bomb.
>
> They paint with kindness and beauty, as if that song
> must be sung, that corpse embraced, those right words said:
> Hiroshima, Nagasaki, Belsen, Dachau, and Vietnam.
>
> And history's heroes are here to be walked upon,
> and *we* are here, beneath their brushstrokes' tread.
> In Iri and Toshi Maruki's painting, the bomb
> is Hiroshima, Nagasaki, Belsen, Dachau, and Vietnam.[24]

Here, as in "Panel I," the Marukis "are painting the bomb," the present progressive insisting on the ongoing nature of their work, even into the future implied in the other refrain, "in Hiroshima, Nagasaki, Belsen, Dachau, and Vietnam," atrocities for which there are and are not yet murals—and a series to which the reader, following the practice of the poet, might add. These lines shift subtly over the course of the poem with the last stanza emphasizing the literal and figurative role that the bomb might play: "In Iri and Toshi Maruki's painting, the bomb / is Hiroshima, Nagasaki, Belsen, Dachau, and Vietnam." Here, the bomb becomes the ur-figure, the template for the series in the subsequent line's refrain. Adding Belsen, Dachau, and Vietnam, recasting their work in a new form, the poem at once acknowledges the Marukis' example and suggests that, learning from their practice, we must carry it forward in whatever medium we know, sestinas or villanelles, literary criticism or cultural studies, ecocriticism, or, as I have suggested here, "risk criticism" and "precautionary reading."

In the *Bulletin*'s "second nuclear age," as in the Marukis' murals, "the bomb" becomes both a reality and a figure, a vehicle for understanding the "hell" that we make for ourselves, in which we are victims and complicit perpetrators of the violence. In fact, tropological wagers proliferate in the murals more generally, as multiplying forms of suffering come to stand analogically for each other, and hell becomes a metaphor for a world at war. Wallace's poems remind us of the capaciousness of the original project, even as they test the limits of the synecdochal expansiveness of "the bomb." Fears of artistic cross-contamination (visual art, documentary film, poetry) come to seem, here, as distractions from the serious work of contesting material contaminations, the ethico-political project in which all of us, artists, critics, activists, must collaborate. Any self-satisfaction we might feel now that the Cold War is over—and with it, seemingly, at least for the moment, the specter of total thermonuclear war—clearly cannot be sustained, for to Hiroshima, Nanking, Auschwitz, Minamata, to Belsen, Dachau, Vietnam, we must add Chernobyl, Deepwater Horizon, Fukushima. The bomb is also Bhopal and Bikini, and perhaps will have been genetically modified organisms and plastics. And the bomb is most certainly climate change. As in the Marukis' work, these are not strictly analogous; rather the series suggests a violence that is pervasive, whether fast or slow, intentional or unintentional.

It might be tempting, now, to see ourselves, inhabitants of the Anthropocene, this new "Eaarth," as in a hell without exit, destined to go down with the planet we mistakenly dubbed a spaceship (as though we were always headed for some other destination, some planet B). But to do so would be to miss the precautionary spirit of the Marukis' project. Taking the long view—the view from posterity that sees such diverse historical acts of violence as Hiroshima, Auschwitz, Minamata, Bhopal, Deepwater Horizon as part of the same thing—can be difficult, but as Toshi said of her own *Hiroshima No Pika*, such depictions are wagered "in the hope that . . . knowing will help keep it from happening again." It is, of course, hubristic to believe that, in discussing such things, one is taking up the mantle of so distinguished a couple as the Marukis. There is no doubt that they were exceptional in their extraordinary talent and absolute dedication to peace. We might, though, inspired by Ronald Wallace, an American, a filmgoer, a poet and scholar, a father, take up the more modest work of being one among many carrying elements of their project forward, with as much beauty and kindness as we can muster.

Notes

INTRODUCTION

1. Andrew McMurray, "The Slow Apocalypse: A Gradualistic Theory of the World's Demise," *Postmodern Culture* 6.3 (1996), http://muse.jhu.edu/login?auth=0&type=summary&url=/journals/postmodern_culture/v006/6.3mcmurry.html.
2. Kim Fortun, *Advocacy after Bhopal: Environmentalism, Disaster, New Global Orders* (Chicago: University of Chicago Press, 2001), 354.
3. Board of Directors, "It Is Five Minutes to Midnight," *Bulletin of the Atomic Scientists* (January–February 2007): 66–71.
4. "'Doomsday Clock' Moves Two Minutes Closer to Midnight," Press Release of the *Bulletin of the Atomic Scientists*, January 17, 2007, http://thebulletin.org/press-release/doomsday-clock-moves-two-minutes-closer-midnight (accessed February 24, 2015).
5. Bruno Latour, "Why Has Critique Run Out of Steam? From Matters of Fact to Matters of Concern," *Critical Inquiry* 30 (Winter 2004): 230.
6. Ibid., 229.
7. Ibid., 238.
8. Ibid., 239.
9. Ibid.
10. Jacques Derrida, "No Apocalypse, Not Now," trans. Catherine Porter and Philip Lewis, *Diacritics* 14 (Summer 1984): 21.
11. Ibid., 23.
12. Ibid.
13. Ibid., 26–27.
14. For a description of these arguments in terms of "weak" and "strong" see Christopher Norris, "'Nuclear Criticism' Ten Years On," *Prose Studies* 17.2 (August 1994): 135.
15. Peter Schwenger, *Letter Bomb: Nuclear Holocaust and the Exploding Word* (Baltimore: Johns Hopkins University Press, 1992), xv.
16. Ibid., xi.
17. William Chaloupka, *Knowing Nukes: The Politics and Culture of the Atom* (Minneapolis: University of Minnesota Press, 1992), 12.
18. Kenneth Knowles Ruthven, *Nuclear Criticism* (Melbourne: Melbourne University Press, 1993), 81.
19. Ibid., 82.
20. Richard Klein, "The Future of Nuclear Criticism," *Yale French Studies* 77 (1990): 77.

21. Ibid., 76.

22. An important exception here is Schwenger's *Letter Bomb*. Describing the "postman of Nagasaki," a man whose back was scarred by flash burns, Schwenger acknowledges that "if Sumiteru uncovers his back at a crowded beach, he is showing the onlookers that nuclear war is not just, as Derrida claims, 'fabulously textual'" (11).

23. Derrida, "No Apocalypse," 23.

24. Schwenger, *Letter Bomb*, xiii.

25. Ruthven, *Nuclear Criticism*, 89.

26. For a useful overview of the more theoretical and humanist/ethical varieties of nuclear criticism see Daniel Cordle, "Cultures of Terror: Nuclear Criticism during and since the Cold War," *Literature Compass* 3.6 (2006): 1186–1199.

27. Daniel Zins, "Seventeen Minutes to Midnight," *Nuclear Texts and Contexts* 7 (Fall 1991): 6. As a kind of transition from nuclear to ecocriticism, this brief essay is remarkable, providing a useful analysis of the ways in which nuclear holocaust is metaphorically and materially linked to environmental holocaust. Indeed, Zins arguably offers a kind of proto-risk criticism of the sort I ultimately advocate.

28. Paul Brians, "Farewell to the First Atomic Age," *Nuclear Texts and Contexts* 8 (Fall 1992): 3.

29. Lawrence Buell, *The Future of Environmental Criticism: Environmental Crisis and Literary Imagination* (Malden, MA: Blackwell, 2005), 6.

30. Ibid., 31.

31. Cordle, "Cultures of Terror," 1191.

32. As someone who has followed ecocriticism closely over the past fifteen years, I am mindful of the oversimplification in this highly telescoped history. Ecocriticism is diverse, and, overall, as Buell usefully reminds us, ecocriticism's relationship to other areas of literary and cultural studies "has been unfolding less as a story of dogged recalcitrance—though there has been some of that—than as a quest for adequate models of inquiry from the plethora of possible alternatives that offer themselves from whatever disciplinary quarter" (*Future of Environmental Criticism*, 10).

33. Christopher Norris, *Uncritical Theory: Postmodernism, Intellectuals, and the Gulf War* (London: Lawrence & Wishart, 1992), 98.

34. Kate Soper, *What Is Nature?* (Oxford: Blackwell, 1995), 151.

35. For an example of a kind of poststructuralist ecocriticism, see Verena Conley's *Ecopolitics: The Environment in Poststructuralist Thought* (New York: Routledge, 1997). For a recent book that reflects insistently on the problems of language, art and ecology, see Timothy Morton's *Ecology without Nature: Rethinking Environmental Aesthetics* (Cambridge: Harvard University Press, 2007). One of the most compelling uses of poststructuralism in considering contemporary nuclear issues is Peter van Wyck's *Signs of Danger: Waste, Trauma, and Nuclear Threat* (Minneapolis: University of Minnesota Press, 2005), from which I draw much inspiration here.

36. See Cordle, "Cultures of Terror." Cordle's book *States of Suspense: The Nuclear Age, Postmodernism, and United States Fiction and Prose* (Manchester: Manchester University Press, 2008) also offers a current example of nuclear

criticism, but here Cordle is focused much more clearly on understanding the Cold War period.

37. Roger Luckhurst, "Nuclear Criticism: Anachronism and Anachorism," *Diacritics* 23.2 (Summer 1993): 90.

38. Ruthven, *Nuclear Criticism*, 91.

39. See, for example, Nelta Edwards, "Radiation, Tobacco, and Illness in Point Hope, Alaska: Approaches to the 'Facts' in Contaminated Communities," and Valerie Kuletz, "The Movement for Environmental Justice in the Pacific Islands," both published in *The Environmental Justice Reader*, ed. Joni Adamson, Mei Mei Evans, and Rachel Stein (Tucson: University of Arizona Press, 2002).

40. I should note here the innovative recent work of Tom Cohen, who, in his "Anecographics," builds, too, on Derrida's "No Apocalypse." Though I find much to admire in Cohen's work, I take issue with his substitute synecdoche for the current environmental crisis, which is "climate change." See Cohen, "Anecographics: Climate Change and 'Late' Deconstruction," in *Impasses of the Post-Global: Theory in the Era of Climate Change*, vol. 2, ed. Henry Sussman (Ann Arbor: Open Humanities Press, 2012), 32–57.

41. Norris, "'Nuclear Criticism," 135.

42. Jeanne X. Kasperson, Roger E. Kasperson, Nick Pidgeon, and Paul Slovic, "The Social Amplification of Risk: Assessing Fifteen Years of Research and Theory," in *The Social Amplification of Risk*, ed. Nick Pidgeon, Roger Kasperson, and Paul Slovic (Cambridge: Cambridge University Press, 2003), 29.

43. This is a point Ursula Heise makes in her very useful overview of risk in *Sense of Place and Sense of Planet* (Oxford: Oxford University Press, 2008), 135.

44. Paul Slovic, "The Perception Gap: Radiation and Risk," *Bulletin of the Atomic Scientists* 68.3 (2012): 73–74.

45. Charles Perrow, "Fukushima and the Inevitability of Accidents," *Bulletin of the Atomic Scientists* 67.6 (2011): 52.

46. Ulrich Beck, "Living in the World Risk Society," *Economy and Society* 35.3 (August 2006): 332.

47. Ulrich Beck, *Ecological Enlightenment: Essays on the Politics of the Risk Society*, trans. Mark A. Ritter (Atlantic Highlands, NJ: Humanities Press, 1995), 61. Emphasis in original.

48. Van Wyck, *Signs of Danger*, 83.

49. Ibid.

50. Ulrich Beck, *World at Risk*, trans. Ciaran Cronin (Cambridge: Polity Press, 2009), 35.

51. Beck, "Living," 345.

52. Van Wyck, *Signs of Danger*, 84.

53. Frances Ferguson, "The Nuclear Sublime," *Diacritics* 14.2 (Summer 1984): 4.

54. Beck, *World at Risk*, 197–198.

55. Van Wyck, *Signs of Danger*, 97.

56. Ibid.

57. See Lawrence Buell, *Writing for an Endangered World: Literature, Culture and Environment in the US and Beyond* (Cambridge: Harvard University Press, 2001), 30–54.

58. Heise, *Sense of Place*, 169.

59. Board of Directors, "It Is Six Minutes to Midnight," Press Release, *Bulletin of the Atomic Scientists* (January 14, 2010), http://turnbacktheclock.org/press-release/it-6-minutes-midnight.

60. See Dipesh Chakrabarty, "The Climate of History: Four Theses," *Critical Inquiry* 35 (Winter 2009): 207.

61. McMurray, "The Slow Apocalypse."

62. Frederick Buell, *From Apocalypse to Way of Life* (New York: Routledge, 2004), 105.

63. Jacques Derrida, "Autoimmunity: Real and Symbolic Suicides—a Dialogue with Jacques Derrida," in *Philosophy in a Time of Terror: Dialogues with Jürgen Habermas and Jacques Derrida*, ed. Giovanna Borradori (Chicago: University of Chicago Press, 2003), 98. Clearly, considering his focus on 9/11, Derrida's interest here is primarily in terror. Beck too has begun in one of his latest books (*World at Risk*) to include terror in risk society, a provocative move that is also, in my view, potentially problematic. While viewing terror as akin to ozone holes or nuclear meltdowns—the unanticipated consequences of modernity—illuminates a Western perspective on such attacks, it also risks collapsing different "risks" that are quite disparate.

64. Van Wyck, *Signs of Danger*, 20.

65. Lydia Millet, *Oh Pure and Radiant Heart* (New York: Soft Skull Press, 2005), 521. This is hardly a novel sentiment. As McMurray points out, it has been anticipated by Baudrillard and R.E.M., among others (see "The Slow Apocalypse," pars. 21–22).

66. Derrida, "No Apocalypse," 207.

67. Ibid., 97, 102.

68. Quoted in Latour, "Critique out of Steam," 226.

69. Ibid., 227.

70. Ibid.

71. Beck, *World at Risk*, 116.

72. "DoD News Briefing—Secretary Rumsfeld and Gen. Myers," Presenter: Secretary of Defense Donald H. Rumsfeld (February 12, 2002). http://www.defense.gov/transcripts/transcript.aspx?transcriptid=2636 .

73. Kerry Whiteside, *Precautionary Politics: Principle and Practice in Confronting Environmental Risk* (Cambridge: MIT Press, 2006), 9.

74. François Ewald, "The Return of Descartes's Malicious Demon: An Outline of a Theory of Precaution," trans. Stephen Utz, in *Embracing Risk*, ed. Tom Baker and Jonathan Simon (Chicago: University of Chicago Press, 2002), 287.

75. Bruno Latour, "Morality and Technology: The End of the Means," trans. Couze Venn, *Theory, Culture and Society* 15.5–6 (2002): 258.

76. Whiteside, *Precautionary Politics*, 27.

77. Ibid.

78. Or they might not. My three-year-old daughter informs me that the song goes "perhaps she'll cry."

79. Slavoj Žižek, *In Defense of Lost Causes* (New York: Verso, 2009), 460.

80. Rob Nixon, *Slow Violence and the Environmentalism of the Poor* (Cambridge: Harvard University Press, 2011).

81. Gay Hawkins, "Plastic Materialities," in *Political Matter: Technoscience,*

Democracy, and Public Life, ed. Bruce Braun and Sarah J. Whatmore (Minneapolis: University of Minnesota Press, 2010), 121.

82. Beck, *World at Risk*, 53.

CHAPTER 1

1. Jacques Derrida, *Archive Fever*, trans. Eric Prenowitz (Chicago: University of Chicago Press, 1998), 18.
2. Robert J. Oppenheimer, "Prospects in the Arts and Sciences," in *The Open Mind* (New York: Simon and Schuster, 1955), 134.
3. Don DeLillo, *Underworld* (New York: Scribner, 1997), 791.
4. Derrida, *Archive Fever*, 3.
5. Ulrich Beck, *Risk Society*, trans. Mark Ritter (London: Sage, 1992), 72.
6. Michel Foucault, *Power/Knowledge: Selected Interviews and Other Writings, 1972–1977*, ed. Colin Gordon, trans. Colin Gordon, Leo Marshall, John Mepham, and Kate Soper (New York: Pantheon, 1972), 128.
7. Beck, *Risk Society*, 83.
8. Ibid., 156.
9. Ibid., 73.
10. Ibid., 72.
11. Hayden White, "Interpretation in History," *New Literary History* 4.2 (Winter 1973): 312.
12. Beck, "Living," 329.
13. D. C. Muecke, *The Compass of Irony* (London: Methuen, 1969), 134.
14. Ibid., 135.
15. To take examples from film, consider *Dr. Strangelove* (1964), *The Atomic Café* (1982), and *War Games* (1983), as a starting point.
16. Beck, *Risk Society*, 11.
17. See, for example, the interview with Oppenheimer included in the documentary film *The Day After Trinity* (1981).
18. Sheila Jasanoff, "Risk in Hindsight: Toward a Politics of Reflection," in *Risk Society and the Culture of Precaution*, ed. Ingo Richter, Sabine Berking, and Ralf Müller-Schmidt (New York: Palgrave Macmillan, 2006), 29–30.
19. Richard Rhodes, *Making the Atomic Bomb* (New York: Simon and Schuster, 1987), 665.
20. Muecke, *The Compass of Irony*, 135.
21. Martin Fackler, "Leak Found in Steel Tank for Water at Fukushima," *New York Times* (June 5, 2013). http://www.nytimes.com/2013/06/06/world/asia/tepco-says-water-at-fukushima-is-contaminated.html?_r=2&.
22. William Deresiewicz, "'I Was There': On Kurt Vonnegut," *Nation* (June 4, 2012). http://www.thenation.com/article/167921/i-was-there-kurt-vonnegut#axzz2bmPyvjpO. Interestingly, Vonnegut's brother, Bernard Vonnegut, was a physicist who worked at one point on cloud seeding. Kurt himself majored in chemistry as an undergraduate.
23. Quoted in Daniel Zins, "Rescuing Science from Technocracy: *Cat's Cradle* and the Play of Apocalypse," *Science Fiction Studies* 3.2 (July 1986): 170.
24. Kurt Vonnegut, *Cat's Cradle* (New York: Dell, 1998), 1.

25. Kurt Vonnegut, *Wampeters, Foma, and Granfalloons* (New York: Dell, 1974), 104.
26. Vonnegut, *Cat's Cradle*, 42–43.
27. Ibid., 43.
28. Beck, *Risk Society*, 25.
29. My reference here is to Zygmunt Bauman, *Liquid Modernity* (Malden, MA: Polity Press, 2000).
30. Vonnegut, *Cat's Cradle*, 56.
31. Ibid., 1.
32. Ken Cooper, "The Whiteness of the Bomb," in *Postmodern Apocalypse: Theory and Practice at the End*, ed. Richard Dellamora (Philadelphia: University of Pennsylvania Press, 1995), 79–106.
33. Toni Cade Bambara, *The Salt Eaters* (New York: Vintage, 1981), 242.
34. Ibid.
35. Vonnegut, *Cat's Cradle*, 247, 11.
36. Bambara, *The Salt Eaters*, 205.
37. Ibid., 209.
38. Vonnegut, *Cat's Cradle*, 11.
39. Ibid., 39.
40. Jonathan Schell, *The Fate of the Earth* and *The Abolition* (Stanford: Stanford University Press, 2000), 105.
41. Ibid., 103.
42. Vonnegut himself indicated that the "innocent" scientist was an "old fashioned" idea. In his "Address to the American Physical Society" in 1969, he insists that "younger scientists are extremely sensitive to the moral implications of all they do" (*Wampeters*, 99). "Any young scientist," Vonnegut asserted, perhaps hopefully, "when asked by the military to create a terror weapon on the order of napalm, is bound to suspect that he may be committing a modern sin" (104).
43. Vonnegut, *Cat's Cradle*, 17.
44. In his "Four Master Tropes," Kenneth Burke argues for the association of irony with the dialectic. See Kenneth Burke, "Four Master Tropes," *Kenyon Review* 3.4 (Autumn 1941): 421–38.
45. Beck, "Living," 331.
46. For a useful overview of the early years of the atomic age in general and of the scientists' movement in particular, see Paul Boyer, *By the Bomb's Early Light* (Chapel Hill: University of North Carolina Press, 1985).
47. D. R. Davies, "Atomic Energy and Personal Responsibility," foreword to *One World or None*, ed. Dexter Masters and Katherine Way (London: Latimer House, 1947), viii.
48. Robert J. Oppenheimer, "Atomic Weapons," *Proceedings of the American Philosophical Society* (January 1946): 7, 9.
49. Ibid., 9.
50. Surely the most in-depth account of the movement is still Alice Kimball Smith, *A Peril and a Hope: The Scientists' Movement in America, 1945–47* (Chicago: University of Chicago Press, 1965).

51. Charles Thorpe, "Violence and the Scientific Vocation," *Theory, Culture & Society* 21.3 (2004): 69.
52. Eugene Rabinowitch, "Five Years After," *Bulletin of the Atomic Scientists* (January 1951): 3.
53. Ibid., 5.
54. Quoted in Thorpe, "Violence," 74.
55. Millet, *Radiant Heart*, 92.
56. Ibid., 38.
57. Derrida, *Archive Fever*, 36.
58. Foucault, *Power/Knowledge*, 129.
59. Margot Norris, "Dividing the Indivisible: The Fissured Story of the Manhattan Project," *Cultural Critique* 35 (Winter 1996–1997): 5–38.
60. Rachel Carson, *Silent Spring* (Boston: Houghton Mifflin, 1962).
61. Millet, *Radiant Heart*, 19.
62. Ibid., 68.
63. Millet, *Radiant Heart*, 38.
64. See, for instance, Leo Szilard, "Can We Avert an Arms Race by an Inspection System?" in Masters and Way, *One World or None*.
65. Barton Bernstein, introduction to Leo Szilard, *"Voice of the Dolphins" and Other Stories*, expanded ed. (Stanford: Stanford University Press, 1992), 15–16.
66. Szilard, *Voice of the Dolphins*.
67. Ibid., 167.
68. Rabinowitch, "Five Years After," 5.
69. Szilard, *Voice of the Dolphins*, 108.
70. Ibid.
71. Vonnegut, *Cat's Cradle*, 245.
72. Szilard, *Voice of the Dolphins*, 53.
73. Millet, *Radiant Heart*, 151.
74. Ibid., 314.
75. Sam Keen, "A Secular Apocalypse," *Bulletin of the Atomic Scientists* (January–February 2007): 30.
76. Millet, *Radiant Heart*, 262.
77. John Snyder, *Prospects of Power: Tragedy, Satire, the Essay, and the Theory of Genre* (Lexington: University Press of Kentucky, 1991), 99.
78. Cathy Caruth, "Afterword: Turning Back to Literature," *PMLA* 125.4 (2010): 1087.
79. Millet, *Radiant Heart*, 215.
80. Ibid., 319.
81. Ibid., 320.
82. Ibid., 63.
83. Ibid., 23, 14.
84. Ibid., 236.
85. Ibid., 521.
86. Frank Kermode, *The Sense of an Ending*, 2nd ed. (New York: Oxford University Press, 2000).
87. *Oh Pure and Radiant Heart* thus arguably suffers from what Daniel

Grausam calls "the narrative problem of non-ending" (318). Reading novels published between the end of the Cold War and 9/11, Grausam identifies a phenomenon that he attributes to the end of the nuclear apocalypse and therefore the "rebirth of historical time, as the specter of mass death is lifted from the planet and temporal horizons are inexorably altered by the return of a seemingly lost relationship to futurity" (311). See Daniel Grausam, "'It Is Only a Statement of the Power of What Comes After': Atomic Nostalgia and the Ends of Postmodernism," *American Literary History* 24.2 (2012): 308–336.

88. Millet, *Radiant Heart*, 174.
89. Ibid., 408.
90. Vonnegut, *Cat's Cradle*, 237.
91. Millet, *Radiant Heart*, 480.
92. Quoted in "'Doomsday Clock' Moves Two Minutes Closer to Midnight" (January 17, 2007) *Bulletin of the Atomic Scientists*. http://www.thebulletin.org/content/media-center/announcements/2007/01/17/doomsday-clock-moves-two-minutes-closer-to-midnight.
93. Millet, *Radiant Heart*, 368.
94. Beck, *Risk Society*, 59.
95. Millet, *Radiant Heart*, 92.
96. Elizabeth DeLoughrey, "Radiation Ecologies and the Wars of Light," *Modern Fiction Studies* 55.3 (Fall 2009): 473.
97. Van Wyck, *Signs of Danger*, 5.
98. David Bradley, *No Place to Hide* (Boston: Little, Brown, 1948), 50.
99. The ironically named *Lucky Dragon*, anchored outside the official exclusion zone, was badly contaminated, with all crew members suffering from radiation poisoning.
100. Bradley, *No Place to Hide*, 104.
101. Jassanof, "Hindsight," 39.
102. Rabinowitch, "Five Years After," 3.
103. Enrico Fermi, "Fermi's Own Story," *Chicago Sun-Times* (November 23, 1952). Included in the U.S. Department of Energy's Report, *The First Reactor* (December 1982): 21. http://www.osti.gov/accomplishments/documents/fullText/ACC0044.pdf.
104. Enrico Fermi, "The Development of the First Chain Reacting Pile," *Proceedings of the American Philosophical Society* 90.1 (January 1946): 24.
105. "Leo Seren, 83; Physicist on the Atomic Bomb Who Turned Pacifist," Obituaries, *Los Angeles Times* (January 11, 2002). http://articles.latimes.com/2002/jan/11/local/me-passings11.2.
106. Jon Anderson, "50 Years after Hiroshima: Nuclear Protesters Collect Their Thoughts," *Chicago Tribune* (August 4, 1995). http://articles.chicagotribune.com/1995-08-04/news/9508040365_1_hiroshima-bomb-uranium-atom.
107. Latour, "Critique Out of Steam," 247.
108. Quoted in ibid.
109. Ibid.
110. Ibid.
111. Latour, "Critique Out of Steam," 248.
112. C. P. Snow, quoted in Zia Mian, "Out of the Nuclear Shadow: Scientists

and the Struggle against the Bomb," *Bulletin of the Atomic Scientists* 71.1 (2015): 60.
113. Fermi, "Fermi's Own Story," 25.
114. Nixon, *Slow Violence*.
115. Millet, *Radiant Heart*, 531.
116. Bambara, *The Salt Eaters*, 245.
117. Ibid., 245–46.
118. Ibid., 246.

CHAPTER 2

1. Lawrence Buell, "Toxic Discourse," *Critical Inquiry* 24.3 (Spring 1998): 642.
2. Nixon, *Slow Violence*, 232.
3. I presented an earlier version of this argument as "Bhopal and the Global (Risk) Village" at a seminar titled "Ecocriticism, Globalization, and Cosmopolitanism" organized by Ursula Heise for the 2009 Association for the Study of Literature and Environment conference in Victoria, British Columbia. I owe a debt to Heise's own book and the essays and excerpts she selected for the seminar participants to read, which have shaped the argument; thanks, too, to other seminar participants for their insights.
4. Bradley, *No Place to Hide*, 163.
5. Ibid., xvii.
6. Schell, *Fate of the Earth*, 46.
7. David Bradford, "We All Live in Bhopal," in *The Bhopal Reader*, ed. Bridget Hanna, Ward Morehouse, and Satinath Sarangi (New York: Apex Press), 283–86.
8. Temik, the brand name for aldicarb, is in the process of being phased out of agricultural use in the United States; the full ban is slated to go into effect by 2018. It was banned in the European Union in 2007.
9. See David Bradford, "We All Live in Bhopal," http://www.eco-action.org/dt/bhopal.html. A slightly different version of the essay is reprinted in Hanna, Morehouse, and Sarangi, *The Bhopal Reader* (283–86).
10. Bradley, *No Place to Hide*, 104.
11. Beck, *World at Risk*, 9–10.
12. Paul de Man, *Allegories of Reading* (New Haven: Yale University Press, 1979), 14.
13. Beck, *World at Risk*, 56, 3.
14. The first baby teeth study in the 1950s and 1960s determined that the above-ground testing was distributing strontium 90 throughout the population. Recent studies have shown an increase in the teeth of U.S. children born following the Chernobyl accident, and presently show significantly higher amounts in children born in the vicinity of nuclear reactors. See http://www.radiation.org/projects/tooth_fairy.html.
15. Heise, *Sense of Place*.
16. Anil Agaral, Juliet Merrifield, and Rajessh Tandon, *No Place to Run: Local*

Realities and Global Issues of the Bhopal Disaster (New Market, TN: Highlander Center; New Delhi: Society for Participatory Research in Asia, 1985), 15.

17. "2013 Mission Statement," ICJB Campaign Posts, March 28, 2013. http://www.bhopal.net/2013-mission-statement/.

18. Beck, *World at Risk*, 98, 96.

19. Ibid., 98.

20. Matthew Packer, "'At the Dead Center of Things' in Don DeLillo's *White Noise*: Mimesis, Violence, and Religious Awe," *Modern Fiction Studies* 51.3 (Fall 2005): 648–66.

21. Criticism on *White Noise* is so voluminous that any attempt to represent it must necessarily be cursory. For work on globalization, see Thomas Peyser, "Globalization in America: The Case of Don DeLillo's *White Noise*," *Clio* 25.3 (1996): 255–271. For whiteness, see Tim Engles, "'Who Are You, Literally': Fantasies of the White Self in *White Noise*," *Modern Fiction Studies* 45.3 (1999): 755–787.

22. Don DeLillo, *White Noise: Text and Criticism*, ed. Mark Osteen (New York: Penguin, 1998), 161.

23. Ibid., 160.

24. Criticism attentive to risk includes Lawrence Buell's analyses of "toxic discourse," Frederick Buell's discussion of apocalyptic satire (in the final chapter of his book *From Apocalypse to Way of Life*), Ursula Heise's explorations of risk and narrative, and, most recently, Susan Mizruchi's discussion of risk and (literary) containment: Susan Mizruchi, "Risk Theory and the Contemporary American Novel," *American Literary History* 22.1 (Spring 2010): 109–135. See also Jennifer Ladino's reading of the novel as evincing an ambivalent nostalgia toward "nature," which complicates the notion that the novel is merely "post-natural": Jennifer Ladino, *Reclaiming Nostalgia: Longing for Nature in American Literature* (Charlottesville: University of Virginia Press, 2012).

25. Buell, *Writing for Endangered World*, 51.

26. DeLillo, *White Noise*, 206.

27. Buell, *From Apocalypse*, 299.

28. Michael Messmer, "Thinking It Through Completely': The Interpretation of Nuclear Culture," *Centennial Review* 32.4 (Fall 1988): 399.

29. DeLillo, *White Noise*, 206.

30. Among the reviews are Diane Johnson's "Conspirators," *New York Times Review of Books* (March 14, 1985); Jayne Anne Phillips's review in *New York Times* (January 13, 1985).

31. Osteen, in DeLillo, *White Noise*, vii. I should note here that Osteen's numbers are fairly low—an Amnesty International study puts the numbers closer to seven to ten thousand instantly, and then some fifteen thousand in subsequent years. And, given the fact that there was no cleanup and the chemicals are persistent, the death toll arguably continues to rise, with some putting it far higher. For alternative numbers see Mike Davis, *Planet of Slums* (London: Verso, 2007), 130.

32. "*Newsweek*," in DeLillo, *White Noise*, 355. In this context, Elise Martucci's acknowledgment (in her citing of these articles in her chapter on *White Noise*) that "the real life toxic event was actually more fantastic and more damaging than DeLillo's fictionalized version" (91) seems like something of an under-

statement. See Elise Martucci, *The Environmental Unconscious in the Fiction of Don DeLillo* (New York: Routledge, 2007).

33. DeLillo, *White Noise*, 281.
34. "*Newsweek*," in DeLillo, *White Noise*, 355.
35. Ibid., 356.
36. Ibid. These articles don't cover a controversy that developed over treatment of the gas affected immediately following the event. Because of uncertainties regarding the composition and operation of the chemicals involved, some physicians wondered whether MIC might break down into cyanide in the body—and thus whether the remedy for cyanide poisoning, sodium thiosulfate, should be administered to victims. One doctor did so with promising results, but this remedy was subsequently withheld because there was no proof of a connection between MIC and cyanide.
37. "*Newsweek*," in DeLillo, *White Noise*, 362.
38. Ibid., 354.
39. Jayne Anne Phillips, "Crowding Out Death," *New York Times* (January 13, 1985).
40. See Mervyn Rothstein, "Theater: A Novelist Faces His Themes on New Ground," *New York Times* (December 20, 1987). http://www.nytimes.com/1987/12/20/theater/theater-a-novelist-faces-his-themes-on-new-ground.html?pagewanted=all&src=pm.
41. Michael Bérubé, "Plot Summary: Motives and Narrative Mechanics in *Underworld* and *White Noise*," in *Approaches to Teaching DeLillo's White Noise*, ed. Tim Engles and John N. Duvall (New York: Modern Language Association, 2006), 141.
42. DeLillo, *White Noise*, 10.
43. For a reading of the supermarket in *White Noise*, see David Alworth, "Supermarket Sociology," *New Literary History* 41 (2010): 301–27.
44. DeLillo, *White Noise*, 131.
45. Agaral, Merrifield, and Tandon, *No Place to Run*, 11. DeLillo's *White Noise* could be just as "prescient" of the U.S. context. As Kim Fortun notes, just subsequent to the accident in Bhopal, there were numerous leaks and spills in Institute and the surroundings areas. On August 11, 1985, "A cloud of chemicals used in the manufacture of aldicarb (Temik) spread over four communities in Institute, West Virginia," and residents referred to subsequent month as "toxic hell month":

> On August 18, a tank truck accident leaked sulfur trioxide, requiring the evacuation of thirty homes. On August 26, a cloud of water and hydrochloric acid leaked and drifted over South Charleston. . . . On September 7, an undisclosed amount of dimethyl disulfide and Larvin leaked from the Institute plant, also without any alarms sounding. On September 8, a leak of methyl mercaptain in nearby Nitro hospitalized three workers. On September 11, the South Charleston plant leaked liquid monomethylamine, but did not report the release until three days later.

See Fortun, *Advocacy after Bhopal*, 61.

46. Sheila Jasanoff, "Bhopal's Trials of Knowledge and Ignorance," *Isis* 98.2 (2007): 348.
47. DeLillo, *White Noise*, 35.
48. Fortun, *Advocacy after Bhopal*, 76–77, quoting from David Weir, *The Bhopal Syndrome* (San Francisco: Center for Investigative Reporting, 1988).
49. DeLillo, *White Noise*, 114.
50. Beck, *World at Risk*, 30.
51. DeLillo, *White Noise*, 174.
52. Ibid., 161–62.
53. John Duvall, "The (Super)Market of Images: Television as Unmediated Mediation in DeLillo's *White Noise*," in DeLillo, *White Noise*, 436.
54. Ibid., 434.
55. Messmer, "Thinking It Through," 404.
56. DeLillo, *White Noise*, 66.
57. Ilan Stavans, Introduction to *Cesar Chavez: An Organizer's Tale* (New York: Penguin, 2008), xxiii.
58. Ibid., xxii.
59. Stavans, *Cesar Chavez*, 200–201. Pesticides were a major focus of the "Wrath of Grapes" campaign, which postdated *White Noise*, but Chávez was pointing to the dangers of pesticides to workers and consumers as early as 1968, when, in a speech later published in *Catholic Worker*, he noted: "I don't eat grapes because I know about these pesticides. You can stop eating grapes for your safety as well as for the boycott." See Richard Jensen and John Hammerback, eds., *The Words of César Chávez* (College Station: Texas A&M University Press, 2002), 33.
60. DeLillo, *White Noise*, 36.
61. Ibid., 161.
62. Ibid., 37–38.
63. Helena María Viramontes, *Under the Feet of Jesus* (New York: Plume, 1995), 50.
64. Ibid., 77.
65. DeLillo, *White Noise*, 170.
66. César Chávez, "The Wrath of Grapes," in *Words of César Chávez*, 132.
67. DeLillo, *White Noise*, 83.
68. David Wilson, "Despite Boycott, Grape Sales Are Up," *New York Times* (November 27, 1988). http://www.nytimes.com/1988/11/27/us/despite-boycott-grape-sales-are-up.html?ref=davidswilson.
69. Agaral, Merrifield, and Tandon, *No Place to Run*, 30.
70. Jasanoff, "Bhopal's Trials," 692.
71. DeLillo, *White Noise*, 66.
72. Ibid., 161–62.
73. Indra Sinha, *Animal's People* (New York: Simon & Schuster), 5.
74. Lawrence Buell, foreword to *Environmental Criticism for the Twenty-First Century*, ed. Stephanie LeMenager, Teresa Shewry, and Ken Hiltner (New York: Routledge, 2011), xv. Though published fairly recently, *Animal's People* has already produced a lively critical conversation. See, for instance, Allison Carruth's "The City Refigured: Environmental Vision in a Transgenic Age," in

LeMenager, Shewry, and Hiltner, *Environmental Criticism*; Heather Snell, "Assessing the Limitations of Laughter in Indra Sinha's Animal's People," *Postcolonial Text* 4.4 (2008): 1–15; and Anthony Carrigan, "'Justice Is on Our Side'? *Animal's People*, Generic Hybridity, and Eco-crime," *Journal of Commonwealth Literature* 47.2 (2012): 159–74. Susie O'Brien has also recently discussed the novel in terms of risk. See Susie O'Brien, "Postcolonial Fiction and the Temporality of Global Risk," Time and Globalization Conference, McMaster University, Workers Arts and Heritage Center, Hamilton, ON, October 19–20, 2012, which is presently available at http://www.academia.edu/2125596/Postcolonial_Fiction_and_the_Temporality_of_Global_Risk.

75. Sinha, *Animal's People*, 14.
76. Ibid., 16.
77. Ibid., 29.
78. Ibid., 65.
79. Nixon, *Slow Violence*, 65.
80. Ramachandra Guha and Juan Martinez-Alvier, *Varieties of Environmentalism: Essays North and South* (Oxford: Oxford University Press, 1997), xxi.
81. Sinha, *Animal's People*, 15.
82. Ibid., 24.
83. Ibid., 7.
84. Ibid., 54.
85. Ibid.
86. Ibid., 69.
87. Ibid., 296.
88. Ibid., 361.
89. Ibid., 360.
90. Beck, *World at Risk*, 99.
91. Sinha, *Animal's People*, 361.
92. Ibid.
93. Ibid.
94. Ibid., 10.
95. Beck, *World at Risk*, 59.
96. Bron Sibree, "The Pen versus Poison," *Courier Mail* (July 7, 2007).
97. Ibid.
98. Indra Sinha, "Bhopal: A Novel Quest for Justice," *Guardian* (October 10, 2007). http://www.theguardian.com/world/2007/oct/10/india-bhopal.
99. Qtd. in Nixon, *Slow Violence*, 48.
100. Sinha, *Animal's People*, 296.
101. Ibid., 248.
102. Ibid.
103. Nixon, *Slow Violence*, 292 n. 5.
104. DeLillo, *White Noise*, 141.
105. Sinha, *Animal's People*, 37.
106. Ibid., 27.
107. Roman Jakobson, "Two Aspects of Language and Two Types of Aphasic Disturbances," in *On Language*, ed. Linda R. Waugh and Monique Monville Burston (Cambridge: Harvard University Press, 1971), 124.

108. Sinha, *Animal's People*, 100.
109. David Lodge, *The Modes of Modern Writing: Metaphor, Metonymy and the Typology of Modern Literature* (London: Edward Arnold, 1977), 75.
110. Sinha, *Animal's People*, 236.
111. Ibid., 229.
112. Quoted in Dominique Lapierre and Javier Moro, *Five Past Midnight in Bhopal* (New York: Warner Books, 2002), 102.
113. Paul Ehrlich, *The Population Bomb* (New York: Ballantine, 1968). Ehrlich acknowledges in the front matter of that book that the phrase is not his originally but rather is taken from a 1954 pamphlet issued by the Hugh Moore Fund.
114. Ibid., 15.
115. Ibid.
116. Of course, even as Ehrlich's anecdote empties the Indian context, making of it a metaphor for global population, the source domain for this metaphor is important—the excesses of population are mostly "theirs"; the solutions "ours." And this is the relationship that "We all live in Bhopal" is deployed to counter.
117. Mizruchi, "Risk Theory," 119.
118. Carson herself, as numerous critics have pointed out, borrowed rhetorical power from the Cold War fears of nuclear fallout, in the case of pesticides, turned back on the U.S. populace.
119. Stavans, *An Organizer's Tale*, 207.
120. Ibid., 207.
121. Michel Foucault, *"Society Must Be Defended"* (London: Picador, 2003), 241.
122. Nixon, *Slow Violence*, 232.
123. Shiv Visvanathan, "Bhopal: The Imagination of a Disaster," *Alternatives* 11 (1986): 154.
124. Ibid.

CHAPTER 3

1. Quoted in B. Julie Johnson, "Animal Patenting," *E: The Environmental Magazine* (April 1994).
2. Whiteside, *Precautionary Politics*, 9.
3. Timothy Morton, *The Ecological Thought* (Cambridge: Harvard University Press, 2012), 86.
4. Freeman Dyson, "Our Biotech Future," in *The Best American Science and Nature Writing*, ed. Jerome Groopman (Boston: Houghton Mifflin, 2008), 64–65.
5. This chapter first appeared as an essay published in *Arizona Quarterly* 67.4 (2011); revised version printed by permission of the Regents of the University of Arizona.
6. For controversy surrounding GM food aid, see Jennifer Clapp, "The Political Economy of Food Aid in an Era of Agricultural Biotechnology," *Global Governance* 11 (2005): 467–85. In 2013, China refused U.S. corn that contained a form of GM corn not approved for import. Chuin-Wei Yap, "China Rejects U.S.

Corn Imports after Finding GMO Strain in Cargoes," *Wall Street Journal* (December 20, 2013). http://online.wsj.com/news/articles/SB10001424052702304866904579269301661883792.

7. Dyson, "Our Biotech Future," 65.

8. Discussing GM food, scientist Lord Robert Winston attempts to discredit activists by granting that it is "a natural human reaction to be wary of scientific progress" (qtd. in Edward Black, "Top TV Scientist Backs GM Crops, but Fears Cloning." *Scotsman* [August 15, 2003]. http://www.bic.searca.org/news/2003/aug/eur/15.html). Though Pope John Paul II opposed GM food, his successor, Pope Benedict XVI, reversed position. See "Genetically Modified Crops Get Vatican's Blessing," *New Scientist* (June 4, 2009). http://www.newscientist.com/article/mg20227114.200-genetically-modified-crops-get-the-vaticans-blessing.html.

9. The argument that genetically modified foods will save the world's hungry is one often made by biotech industry. Indeed, capitalizing on fears of environmental scarcity, the Monsanto corporation has recently produced a glossy advertising campaign, which has appeared in the *New Yorker* magazine among other places, titled "How Can We Squeeze More Food From a Raindrop." See http://www.monsanto.com/pdf/sustainability/advertisement.pdf. For an argument that some GM foods might cause cancer, see Jeffrey M. Smith's descriptions of the testing of a lectin-producing potato by Arpad Pusztai, who found lower growth rates, liver problems, and cell proliferation in rats fed the potatoes. According to Smith, Pusztai's results "strongly suggested that the GM foods already approved and being eaten by hundreds of millions of people every day might be creating similar health problems in people, especially in children." See Jeffrey Smith, *Seeds of Deception: Exposing Industry and Government Lies about the Safety of the Genetically Engineered Foods You're Eating* (Fairfield, IA: Yes! Books, 2003), 13.

10. See Ian Hacking, "Our Neo-Cartesian Bodies in Parts," *Critical Inquiry* 34 (Autumn 2007): 78–105; and Donna Haraway, *Modest_Witness@Second_Millennium* (New York: Routledge, 1997).

11. Susan McHugh is correct, however, in pointing out that the bulk of this work concerns transgenic animals and human beings rather than plants. See Susan McHugh, "Flora, Not Fauna: GM Culture and Agriculture," *Literature and Medicine* 26.1 (Spring 2007): 25–54.

12. Beck, *World at Risk*, 83.

13. Ibid., 35.

14. Claire Hope Cummings, *Uncertain Peril: Genetic Engineering and the Future of Seeds* (Boston: Beacon Press, 2008).

15. The Council subsequently published an anthology, edited by Leon Kass, titled *Being Human: Core Readings in the Humanities* (New York: Norton, 2004). For provocative readings of Victorian and Romantic literature in light of contemporary biotechnology, see Jay Clayton, "Victorian Chimeras, or, What Literature Can Contribute to Genetics Policy Today," *New Literary History* 38 (2007): 569–91; and Anne-Lise François, "'Oh Happy Living Things': Frankenfoods and the Bounds of Wordsworthian Natural Piety," *Diacritics* 33.2 (Summer 2003): 42–70.

16. Ward Moore, *Greener Than You Think* (Rockville, MD: Wildside Press, originally published 1947), 11, 51.

17. Qtd. in Robert Streiffer and Thomas Hedemann, "The Political Import of Intrinsic Objections to Genetically Engineered Food," *Journal of Agricultural and Environmental Ethics* 18.2 (2005): 192.

18. *Pandora's Picnic Basket* is the title of a book (published by Oxford University Press in 2000) by Alan McHughen, himself a scientist who has worked with GMOs.

19. Qtd. In Janet Bainbridge, "The Use of Substantial Equivalence in the Risk Assessment of GM Food," May 16, 2001. www.royalsociety.org. Second emphasis added.

20. Eve Tavor Bannet, "Analogy as Translation: Wittgenstein, Derrida, and the Law of Language," *New Literary History* 28 (1997): 655.

21. Ibid., 656.

22. Erik Millstone, Eric Brunner, and Sue Mayer, "Beyond 'Substantial Equivalence,'" *Nature* 401 (October 7, 1999): 526.

23. Harry A. Kuiper, Gijs A. Kleter, Hub P. J. M. Noteborn, and Esther J. Kok, "Substantial Equivalence—an Appropriate Paradigm for the Safety Assessment of Genetically Modified Foods?" *Toxicology* 181–182 (2002): 427. Emphasis in the original.

24. Ursula LeGuin, introduction to *The Left Hand of Darkness* (New York: Ace, 1976).

25. Buell, *Writing for Endangered World*, 48.

26. Indeed, the Bt crops like NewLeaf™ potatoes can be read as Monsanto's ironic extension of the final chapter of *Silent Spring*, in which Carson suggests that pest control in the future will be biological rather than chemical, and that Bt might be a good alternative to DDT and other chemical pesticides. At present, Bt as a spray is allowed in organic farming, and organic farmers warn that incorporating this gene into plants may accelerate insect resistance, a possibility that scientists and life sciences corporations acknowledge is inevitable.

27. Buell, *Writing for Endangered World*, 47–48.

28. Beck, *World at Risk*, 12.

29. Ruth Ozeki, *All Over Creation* (New York: Viking, 2003), 291.

30. Regulation of GM crops is Canada is much the same as in the United States. Indeed, one of the most widely publicized controversies surrounding patent policing was the case of Percy Schmeiser, a Canadian canola farmer who was sued by Monsanto for patent infringement. For a literary treatment of this case, see Annabel Soutar's play *Seeds* (Vancouver: Talon Books, 2012).

31. Ozeki, *All Over Creation*, 92.

32. Ibid., 89.

33. Buell, *Writing for Endangered World*, 51.

34. Beck, *World at Risk*, 50.

35. David Palumbo-Liu, "Rational and Irrational Choices: Form, Affect, and Ethics," in *Minor Transnationalism*, ed. Francoise Lionnet and Shu-mei Shih (Durham: Duke University Press, 2005), 54.

36. As Palumbo-Liu in "Rational and Irrational Choices" notes, this seeming

incoherence characterizes *My Year of Meats* as well, though he provides an elegant analysis of it as an intentional political strategy. For another critique of Ozeki's first novel on this front, see Monica Chiu, "Postnational Globalization and (En)Gendered Meat Production in Ruth L. Ozeki's *My Year of Meats*," *LIT: Literature, Interpretation, Theory* 12.1 (2001): 99–128.

37. Judith Beth Cohen, "Bad Seeds," review of *All Over Creation*, *Women's Review of Books* 20.8 (May 2003): 6.

38. Streiffer and Hedmanm, "Political Import," 193.

39. Michael Pollan, "Playing God in the Garden," *New York Times Magazine* (October 25, 1998). http://www.nytimes.com/1998/10/25/magazine/playing-god-in-the-garden.html.

40. Ozeki, *All Over Creation*, 85.

41. Ibid., 105.

42. Ibid., 106.

43. Ibid., 173.

44. Haraway, *Modest Witness*, 60.

45. Ibid., 61.

46. Ozeki, *All Over Creation*, 124. Geek is referring to the use of flounder genes to improve cold-tolerance.

47. Ibid., 67.

48. See Ursula Heise, "Ecocriticism and the Transnational Turn in American Studies," *American Literary History* 20.1–2 (Spring–Summer 2008): 399.

49. For a classic analysis of the "ecological" side of imperialism, see Alfred Crosby's *Ecological Imperialism*, 2nd ed. (New York: Cambridge University Press, 2004).

50. Ozeki, *All Over Creation*, 301.

51. Vandana Shiva, *Stolen Harvest: The Hijacking of the Global Food Supply* (Cambridge, MA: South End Press, 2000), 82.

52. Ozeki, *All Over Creation*, 267.

53. Ibid., 124.

54. For a provocative reading of the interplay of sexual politics and seed politics in the novel, see Rachel Stein, "Bad Seed," in *Postcolonial Green: Environmental Politics and World Narratives*, ed. Bonnie Roos and Alex Hunt (Charlottesville: University of Virginia Press, 2010), 177–93.

55. One could certainly read this unlikely alliance of political positions more affirmatively. In an essay on finding common ground in religious and secular humanist views of the environment, Jennifer Ladino offers *All Over Creation* as an example of an imagined "green culture of life" that "combines compassion for others, respect for natural and cultural diversity, and a dedication to rectifying geopolitical inequality" (153). See Ladino, "Unlikely Alliances: Notes on a Green Culture of Life," *Journal of Religion and Society* (2008): 146–58.

56. Michael Hardt and Antonio Negri, *Multitude* (New York: Penguin, 2004), 183–84.

57. Morton, *The Ecological Thought*, 86.

58. Donna Haraway, *The Companion Species Manifesto* (Chicago: Prickly Paradigm Press, 2003), 11.

59. See Arjun Appadurai, *Modernity at Large* (Minneapolis: University of Minnesota Press, 1996), in which he comments on the "hijack of culture by literary studies" (51).

60. Ronald Herring, "Stealth Seeds: Bioproperty, Biosafety, Biopolitics," *Journal of Development Studies* 43.1 (January 2007): 135.

61. See Ronald Herring, "Why Did 'Operation Cremate Monsanto' Fail? Science and Class in India's Great Terminator-Technology Hoax," *Critical Asian Studies* 38.4 (2006): 467–93. Ozeki, in fact, does depict the "termination" of the Terminator in the novel, acknowledging that this technology is not commercialized, though her character "Geek" suggests that, like its Hollywood namesake, "Terminator will be back" (Ozeki, 399).

62. Herring suggests that this is by definition impossible, since a plant engineered to have sterile seeds could not reproduce or cross-pollinate. Nevertheless, other GM plants have mixed with non-GM plants in ways that were not entirely predictable.

63. Herring, "Stealth Seeds," 146.

64. *World According to Monsanto,* dir. Marie-Monique Robin, English edition (Yes! Books, 2008). DVD.

65. Hélène Cixous, "Laugh of the Medusa," *Signs* 1.4 (Summer 1976): 885.

66. Donna Haraway, *Simians, Cyborgs, and Women* (New York: Routledge, 1991), 151.

67. Žižek, *Lost Causes*, 446.

68. Ibid., 428.

69. Ursula Heise, while acknowledging that Lloyd's fundamentalism "makes him something more complex than a simple authorial mouthpiece," suggests that the "reader is quite obviously invited to sympathize" with the Seeds (398–99). And Cohen, noting the ambiguity in the novel's stance toward the Seeds' "self-righteousness," also suggests that they appear to be "our only hope" (6).

70. For a thoughtful reflection on the contemporary agricultural industrial complex, see Susan Willis, "The Forensics of Spinach," *South Atlantic Quarterly* 107.2 (Spring 2008): 355–72.

71. Ozeki, *All Over Creation*, 246.

72. Ibid., 220.

73. Dan Charles, "And the Winner of the World Food Prize Is . . . the Man from Monsanto," NPR (June 19, 2013). http://www.npr.org/blogs/the-salt/2013/06/19/193447482/and-the-winner-of-the-world-food-prize-is-the-man-from-monsanto; Mike Adams, "Monsanto Voted Most Evil Corporation of the Year," *Natural News* (January 10, 2011). http://www.naturalnews.com/030967_Monsanto_evil.html.

74. Herring, "Why Did 'Operation Cremate Monsanto' Fail?," 475. For a recent version of this critique see Marie-Monique Robin's documentary film *The World According to Monsanto* (2008).

75. For versions of this argument, see Henry I. Miller, Gregory Conko, and Drew L. Kershen, "Why Spurning Food Biotech Has Become a Liability," *Nature Biotechnology* 24.9 (September 2006): 1075–77; and Wojciech K. Kaniewski and Peter E. Thomas, "The Potato Story," *AgBioForum* 7.1–2 (2004): 41–46.

76. Ozeki, *All Over Creation*, 219.

77. Ibid., 98.
78. Ibid., 411.
79. Ibid., 344.
80. See Ingo Potrykus, "Golden Rice and Beyond," *Plant Physiology* 125 (March 2001): 1157–61.
81. Monsanto, "Our Commitment to Sustainable Agriculture." http://www.monsanto.com/whoweare/Pages/our-commitment-to-sustainable-agriculture.aspx.
82. *Body 2.0: Creating a World That Can Feed Itself.* Zeitgeist Google Partner Forum (2008). Available at http://www.youtube.com/watch?v=I9I1IkbcHNE . The video is also available at Michael Pollan's website: http://michaelpollan.com/videos/michael-pollan-at-googles-zeitgeist-forum/.
83. Allison Carruth, *Global Appetites* (Cambridge: Cambridge University Press, 2013), 117.
84. William Conlogue, quoted in Carruth, *Global Appetites*, 117.
85. Ross Andersen, "After 4 Years, Checking Up on the Svalbard Global Seed Vault," *Atlantic* (February 28, 2012). http://www.theatlantic.com/technology/archive/2012/02/after-4-years-checking-up-on-the-svalbard-global-seed-vault/253458/.
86. Josh McHugh, "Google versus Evil," *Wired*. http://www.wired.com/wired/archive/11.01/google_pr.html.
87. Quoted in Brian Handwerk, "Doomsday Seed Vault Safeguards Our Food Supply," *National Geographic Magazine* (July 2, 2012). http://news.nationalgeographic.com/news/pictures/2012/07/120702-svalbard-doomsday-seed-vault-food-supply/.
88. Ozeki, *All Over Creation*, 127. On Monsanto's precision technology see Matthew Perrone, "Monsanto Loss Widens, Announces $930M Acquisition," *Seattle PI* (October 2, 2013).
89. Ozeki, *All Over Creation*, 410.
90. Ibid., 123.
91. Ibid., 125.
92. GRAIN, "Faults in the Vault: Not Everyone is Celebrating Svalbard" (February 2008). http://www.grain.org/fr/article/entries/181-faults-in-the-vault-not-everyone-is-celebrating-svalbard.
93. Ozeki, *All Over Creation*, 357.
94. Ibid., 161.
95. Ibid., 171.
96. Ibid.
97. Ruth Ozeki, "Seeds of Our Stories," *Moebius* 6.1 (2008): 9.
98. Ibid.
99. Ibid., 10.
100. McHugh, "Flora, Not Fauna," 37.
101. Beck, *World at Risk*, 33.
102. Ibid., 95, 97.
103. Ron Charles, "Starch-Free Protest," review of *All Over Creation*, *Christian Science Monitor* (March 13, 2003). Posted on *Ruth Ozeki's Web World*. http://www.ruthozeki.com/reviews/csm.html.
104. Beck, *World at Risk*, 97.

105. Ozeki, "Seeds," 25.
106. Beck, *World at Risk*, 111.
107. Greg Garrard, "Nature Cures? or How to Police Analogies of Personal and Ecological Health," *ISLE: Interdisciplinary Studies in Literature and Environment* 19.3 (Summer 2012): 497.
108. Quoted in Garrard, "Nature Cures," 497.
109. Bainbridge, "Use of Substantial Equivalence."
110. Whiteside, *Precautionary Politics*, 27.

CHAPTER 4

1. Dhananjay Mahapatra, "Threat of Plastic Bags BIGGER than the Atom Bomb: SC," *Times of India* (May 7, 2012). http://timesofindia.indiatimes.com/home/environment/pollution/Threat-of-plastic-bags-bigger-than-atom-bomb-SC/articleshow/13031361.cms.
2. Timothy Clark, "Toward a Deconstructive Environmental Criticism," *Oxford Literary Review* 30.1 (2008): 49.
3. Jacques Derrida, "Biodegradables: Seven Diary Fragments," trans. Peggy Kamuf, *Critical Inquiry* 15.4 (Summer 1989): 816–19.
4. Francisco Vara-Orta, "A Reservoir Goes Undercover," *Los Angeles Times* (June 10, 2008). http://articles.latimes.com/2008/jun/10/local/me-balls10.
5. Signs point to no. See Jenna Bilbrey, "BPA-Free Plastic Containers May Be Just as Hazardous," *Scientific American* (August 11, 2014). http://www.scientificamerican.com/article/bpa-free-plastic-containers-may-be-just-as-hazardous/.
6. Marty Adams, "A Letter from Marty Adams, Director of Water Quality & Operations," posted June 25, 2008. Silver Lake Reservoir Conservancy website. http://www.silverlakereservoirs.org/news-and-events/bird-balls-on-ivanhoe-reservoir-are-non-toxic-questions-about-ivanhoe-bird-ball-safety-and-dwps-intentions/.
7. "HDPE Pipe for Potable Water Applications," report by the Pacific Northwest Pollution Prevention Resource Center, February 2010. http://www.google.com/url?sa=t&rct=j&q=&esrc=s&source=web&cd=9&ved=0CGcQFjAI&url=http%3A%2F%2Fwww.pprc.org%2Fresearch%2FrapidresDocs%2FPPPRC_HDPE_Water_Pipe_Safety_FINAL.pdf&ei=z9CxUsWcHsTgyQG91oGoAw&usg=AFQjCNEwwrFvujtQwT0FXB1rgqCu3wLQKg&bvm=bv.58187178,d.aWc.
8. Susan Freinkel, *Plastic: A Toxic Love Story* (New York: Houghton Mifflin Harcourt, 2011), 95.
9. Beck, "Living," 345.
10. See Stacy Alaimo, "States of Suspension: Trans-corporeality At Sea," *ISLE: Interdisciplinary Studies in Literature and Environment* 19.3 (Summer 2012): 486; and Patricia Yaeger, "Editor's Column: Sea Trash, Dark Pools, and the Tragedy of the Commons," *PMLA* 125.3 (May 2010): 527.
11. Veronique de Turenne, "Those Reservoir Balls—Are They Safe?" *Los Angeles Times* (June 10, 2008). http://latimesblogs.latimes.com/lanow/2008/06/franciscos-post.html.

12. Derrida, "Biodegradables," 866.
13. Tze-Yin Teo, "Responsibility, Biodegradability," *Oxford Literary Review* 32.1 (2010): 98.
14. Derrida, "Biodegradables," 814.
15. Van Wyck, *Signs of Danger*, 5.
16. Timothy Morton, *Hyperobjects: Philosophy and Ecology after the End of the World* (Minneapolis: University of Minnesota Press, 2013), 5.
17. Freinkel, *Plastic*, 6.
18. NOAA, "How Big Is the 'Great Pacific Garbage Patch'? Science vs. Myth." http://response.restoration.noaa.gov/about/media/how-big-great-pacific-garbage-patch-science-vs-myth.html.
19. DeLillo, *White Noise*, 280.
20. Jody Roberts, "Reflections of an Unrepentant Plastiphobe," in *Accumulation: The Material Politics of Plastic*, ed. Jennifer Gabrys, Gay Hawkins, and Mike Michael (London: Routledge, 2014), 128.
21. Max Liboiron, "Plasticizers: A Twenty-First Century Miasma," in Gabrys, Hawkins, and Michael, *Accumulation*, 135.
22. Freinkel, *Plastic*, 137.
23. Alaimo, "States of Suspension," 487.
24. Ibid., 486.
25. Paul de Man, "Anthropomorphism and Trope in Lyric," in *The Rhetoric of Romanticism* (New York: Columbia University Press, 1984), 241.
26. Ibid.
27. Jane Bennett, *Vibrant Matter: A Political Ecology of Things* (Durham: Duke University Press, 2010), xvi.
28. Ibid., 25.
29. I take Barbara Johnson's point that anthropomorphism and personification are generally "figures of being, not address," and therefore "what matters in them is their predicates, not their voices" (18). In this definition, the phrase "letting plastic bags have their say" is perhaps a form of anthropomorphism, but ventriloquizing their actual speech may be closer to prosopopoeia, a concept she discusses in conjunction with a talking Barbie. See Johnson, *Persons and Things* (Cambridge: Harvard University Press, 2010).
30. Many a critical debate has been waged on the subtle differences among these related strategies (anthropomorphism, personification, prosopopoeia). Though some of the examples that I cite here are perhaps more akin to the latter, I prefer to retain the term "anthropomorphism" precisely for its suffix (-morphism), which resonates with the "plastic" qualities of plastic in particular.
31. Timothy Morton, "An Object-Oriented Defense of Poetry," *New Literary History* 43.2 (Spring 2012): 221.
32. Serenella Iovino and Serpil Oppermann, "Theorizing Material Ecocriticism: A Diptych," *ISLE: Interdisciplinary Studies in Literature and Environment* 19.3 (Summer 2012): 459.
33. Ibid., 461.
34. Yaeger, "Editor's Column: Sea Trash," 529.
35. Bennett, *Vibrant*, 121.

36. Hawkins, "Plastic Materialities," 120.
37. Ibid., 128, 121.
38. Gay Hawkins, *The Ethics of Waste: How We Relate to Rubbish* (New York: Rowman & Littlefield, 2005), 23.
39. Heal the Bay, *The Majestic Plastic Bag*, dir. Jeremy Konner, 2010.
40. "Jeremy Irons Narrates Heal the Bay Mockumentary: 'The Majestic Plastic Bag.'" http://jeremyirons.com/jeremy-irons-narrates-heal-the-bay-mockumentary-the-majestic-plastic-bag.
41. Patricia Yaeger, "Editor's Column: The Death of Nature and the Apotheosis of Trash; or, Rubbish Ecology," *PMLA* 123.2 (March 2008): 332.
42. Quoted in Hawkins, "Plastic Materialities," 134.
43. In this way, the short recalls Albert Lamorisse's 1956 short film *The Red Balloon*, though clearly in that earlier film the balloon is much more animated and motivated. It indeed is begging the boy-protagonist to play with it.
44. Jonathan Lamb, *The Things Things Say* (Princeton: Princeton University Press, 2011), xvi.
45. James Palmer, "A Review of 'Plastic Bag,'" *Psychological Perspectives* 54.2 (2011): 244.
46. De Man, "Anthropomorphism," 241.
47. Derrida, "Biodegradables," 824.
48. Lamb, *The Things Things Say*, xi.
49. Ibid.
50. Alfred Lord Tennyson, "Tithonus," *Norton Anthology of English Literature*, 5th ed., vol. 2, ed. M. H. Abrams (New York: Norton, 1986), 1110.
51. Bill Brown, "How to Do Things with Things (a Toy Story)," *Critical Inquiry* 24.4 (Summer 1998): 947.
52. Ulrich Beck, *World Risk Society*, trans. Ciaran Cronin (Cambridge: Polity, 1999), 45.
53. Rachel Hope Allison, *I'm Not a Plastic Bag: A Graphic Novel* (Los Angeles: Archaia, 2012).
54. Aaron Long, "Archaia Teams with JeffCorwinConnect for i'm not a plastic bag," Comicuriosity (February 23, 2012). http://www.comicosity.com/archaia-teams-with-jeffcorwinconnect-for-i-am-not-a-plastic-bag/.
55. This section draws in part on a reading of the novel that I perform in a previously published essay, "A Bizarre Ecology: The Nature of Denatured Nature," *ISLE: Interdisciplinary Studies in Literature and Environment* 7.2 (2000): 137–53.
56. Caroline Rody, "Impossible Voices: Ethnic Postmodern Narration in Toni Morrison's *Jazz* and Karen Tei Yamashita's *Through the Arc of the Rain Forest*," *Contemporary Literature* 41.4 (Winter 2000): 629.
57. Karen Tei Yamashita, *Through the Arc of the Rain Forest* (Minneapolis: Coffee House Press, 1990), 202.
58. Myra Hird, "Knowing Waste: Toward an Inhuman Epistemology," *Social Epistemology* 26.3–4 (2012): 465.
59. Heise, *Sense of Place*, 111.
60. Yamashita, *Rain Forest*, 106.
61. Ibid., 5.

62. Heise, *Sense of Place*, 112.
63. Yamashita, *Rain Forest*, 142.
64. Ibid., 168.
65. Ibid., 167, 142.
66. Ibid., 144.
67. Ibid., 96.
68. Ibid., 150.
69. Ibid., 160.
70. Quoted in Van Wyck, *Signs*, 6.
71. Yamashita, *Rain Forest*, 160.
72. Ibid., 203.
73. Ibid., 148.
74. Ibid., 203.
75. Ibid., 212.
76. Yaeger, "Editor's Column: Sea Trash," 530.
77. Bruno Latour, *Politics of Nature: How to Bring the Sciences into Democracy* (Cambridge: Harvard University Press, 2004), 80.
78. Gwyneth Dickey Zaikab, "Marine Microbes Digest Plastic," *Nature* (March 28, 2011). http://www.nature.com/news/2011/110328/full/news.2011.191.html.
79. Damien Gayle, "Could a Plastic-Eating Fungi Save the World from Biggest Man-Made Environmental Catastrophe?" *Daily Mail* (May 18, 2012). http://www.dailymail.co.uk/sciencetech/article-2146224/Could-fungi-break-plastic-stop-modern-scourge.html.
80. Derrida, "Biodegradables," 873.
81. Yamashita, *Rain Forest*, 212.
82. DeLillo, *White Noise*, 160.
83. Quoted in Zaikab, "Marine Microbes Digest Plastic."
84. Morton, "Object-Oriented Defense," 207.
85. Gay Hawkins, "Made to Be Wasted: PET and the Topologies of Disposability," in Gabrys, Gay Hawkins, and Michael, *Accumulation*, 49–67.
86. Stacy Alaimo, *Bodily Natures: Science, Environment, and the Material Self* (Bloomington: Indiana University Press, 2010).
87. "Hail," copyright © 2013 by Adam Dickinson. Reproduced with permission from House of Anansi Press, Toronto. www.houseofanansi.com.
88. Bernadette Bensaude Vincent, "Plastics, Materials and Dreams of Dematerialization," in Gabrys, Hawkins, and Michael, *Accumulation*, 23.
89. Roland Barthes, *Mythologies*, trans. Annette Lavers (New York: Hill and Wang, 1972), 97–98.
90. Chelsea M. Rochman, Mark Anthony Browne, Benjamin S. Halpern, Brian T. Hentschel, Eunha Oh, Hrissi K. Karapanagioti, Lorena M. Rios-Mendoza, Hideshige Takada, Swee Teh, adn Richard C. Thompson, "Classify Plastic Waste as Hazardous," *Nature* 494 (February 14, 2013): 169. http://www.nature.com/nature/journal/v494/n7436/full/494169a.html?WT.ec_id=NATURE-20130214.
91. Ibid., 170.
92. Ibid., 171.

93. Ibid.
94. DeLillo, *Underworld*, 810.
95. United States Environmental Protection Agency, "Wastes—Resource Conservation—Common Wastes & Material: Plastics." http://www.epa.gov/osw/conserve/materials/plastics.htm.
96. Orange Products, "Company History." http://orangeprecisionballs.com/about-us/company-history.
97. Vara-Orta, "A Reservoir Goes Undercover."
98. Yamashita, *Rain Forest*, 142.
99. See Beth Terry's *My Plastic Free Life*. http://myplasticfreelife.com/.
100. Iovino and Opperman, "Theorizing Material Ecocriticism," 461.
101. Barthes, *Mythologies*, 98.

CHAPTER 5

1. Morton, *Hyperobjects*, 103.
2. Ulrich Beck, "Interview / Ulrich Beck: System of Organized Irresponsibility behind the Fukushima Crisis," interview with Hirohito Ohno, *Asahi Shimbun* (July 6, 2011). http://ajw.asahi.com/article/0311disaster/analysis_opinion/AJ201107063167.
3. Terry Tempest Williams, *Refuge: An Unnatural History of Family and Place* (New York: Vintage, 1991), 140.
4. Ibid., 286.
5. Bill McKibben, *The End of Nature* (New York: Random House, 1989), 47.
6. Morton, *Hyperobjects*, 103.
7. To say that nuclear emits "no" greenhouse gases is to ignore the entire nuclear cycle, which would include mining, processing, transport, and disposal.
8. Bill McKibben, *Eaarth: Making a Life on a Tough New Planet* (Toronto: Knopf Canada, 2010), 57.
9. Ibid.
10. Though the organization "Environmentalists for Nuclear Energy Canada," of which Patrick Moore is honorary chairman, might read as comically akin to the "Poets and Painters for Immediate Nuclear War," which Vonnegut imagines as the organization appropriate to his nihilist character Krebbs, this seems to be just the sort of "real-life satire" that Beck imagines as appropriate to risk society.
11. See the page on the website "About NIE." http://www.nei.org/About-NEI. Emphasis added.
12. Suzanne Goldenberg, "Climate Change Is Clear and Present Danger, Says Landmark US Report," *Guardian* (May 4, 2014). "http://www.theguardian.com/environment/2014/may/04/climate-change-present-us-national-assessment.
13. Erik Swyngedouw, "Apocalypse Forever? Post-political Populism and the Spectre of Climate Change," *Theory Culture and Society* 27.2–3 (2010): 213–232.

14. Sikina Jinnah, "Climate Change Bandwagoning: The Impacts of Strategic Linkages on Regime Design, Maintenance, and Death," *Global Environmental Politics* 11.3 (August 2011): 4.

15. Swyngedouw, "Apocalypse Forever," 225.

16. Cohen, "Anecographics," 39.

17. Bryan Walsh, "How to Win the War on Global Warming," *Time* (April 17, 2008). http://content.time.com/time/specials/2007/article/0,28804,1730759_1731383_1731363,00.html.

18. This was the title of George Monbiot's 2011 article in the *Guardian*. See Monbiot, "Why Fukushima Made Me Stop Worrying and Love Nuclear Power," *Guardian* (March 21, 2011). http://www.theguardian.com/commentisfree/2011/mar/21/pro-nuclear-japan-fukushima?commentpage=1.

19. See Richard Black, "Fukushima: As Bad as Chernobyl?" *BBC News* (April 12, 2011). http://www.bbc.co.uk/news/science-environment-13048916.

20. See "John Gummer: Beef Eater," *BBC News: World Edition* (October 11, 2000). http://news.bbc.co.uk/2/hi/uk_news/369625.stm.

21. "Japan Reactors at 'High Risk' for More Leaks." *CBC News*, March 15, 2011. http://news.aol.ca/2011/03/15/japan-reactors-at-high-risk-for-more-leaks-pm/. A video of the strawberry-eating incident is available at http://en.rian.ru/video/20110413/163514291.html.

22. See Beck, "Interview"; and IAEA's website: http://www.iaea.org/newscenter/focus/chernobyl/faqs.shtml.

23. Charles Perrow, "Nuclear Denial: From Hiroshima to Fukushima," *Bulletin of the Atomic Scientists* 69.5 (2013): 64.

24. Ibid., 57.

25. Jacob Darwin Hamlin, "Fukushima and the Motifs of Nuclear History," *Environmental History* 17 (April 2012): 287.

26. Ibid.

27. Ibid., 295.

28. Cohen, "Anecographics," 45.

29. Slavoj Žižek, *First as Tragedy, Then as Farce* (London: Verso, 2009), 87.

30. Rabinowitch, "Five Years After," 3.

31. Canada was not, of course, the only supplier of uranium for the Manhattan Project; a greater percentage originated in what was then the Belgian Congo.

32. Déline refers to the location of the community, at the headwaters of Great Bear River. The area is populated primarily by Sahtu Dene people. Port Radium is some distance from Déline proper, across the lake.

33. "Vanishing present" comes from the subtitle to her 1999 book, *A Critique of Postcolonial Reason: Toward a History of the Vanishing Present* (Cambridge: Harvard University Press, 1999).

34. Peter C. van Wyck, *The Highway of the Atom* (Montreal: McGill-Queen's University Press, 2010), 49.

35. Nixon, *Slow Violence*, 6.

36. Beck, *World at Risk*, 29.

37. Peter Blow, *Village of Widows* (Toronto: Kineticvideo, 1998), 52 min., video.

38. Indeed, as van Wyck reports, at least one critic alleged that the footage

was from the funeral of a victim of a truck accident, though van Wyck notes that this critic, Walter Keyes, was himself not an unbiased viewer (van Wyck, *Highway*, 187).

39. Van Wyck, *Highway of the Atom*, 48.

40. Ibid., 41.

41. Capitalization of the LaBine brothers' names differs from source to source ("Labine" or "LaBine"). I have adopted the spelling in the most recent edition of Clements's *Burning Vision* for consistency.

42. Jill Godmilow, "How Real Is the Reality in Documentary Film?" interview with Ann-Louise Shapiro, *History and Theory* 36.4 (December 1997): 83.

43. Ibid., 84.

44. Ibid.

45. Theresa May, "Kneading Marie Clements' *Burning Vision*," *Canadian Theatre Review* 144 (Fall 2010): 5.

46. Marie Clements, *Burning Vision* (Vancouver: Talon Books, 2003), 19.

47. As Peter van Wyck reports, it was Gilbert who actually staked the original claim. Though he was traveling with another prospector, it was Charlie St. Paul (apparently suffering from snow blindness), not LaBine's brother (see van Wyck, *Highway of the Atom*, 195).

48. Clements, *Burning Vision*, 85.

49. Beck, *World at Risk*, 197–98.

50. Clements, *Burning Vision*, 115.

51. Ibid.

52. Ibid., 117.

53. Ibid., 110.

54. Ibid., 114.

55. Ibid., 115.

56. Ibid., 118.

57. Ewald, "Return of Descartes's Demon," 287.

58. Ibid.

59. Ibid., 288.

60. Clements, *Burning Vision*, 23.

61. Ibid.

62. Ibid., 35.

63. Ibid., 108.

64. Ibid., 119.

65. Ibid., 75.

66. Theresa May, "This Is My Neighborhood," keynote address at "Staging Sustainability" conference, York University, April 20, 2011.

67. Clements, *Burning Vision*, 99.

68. Ibid., 95.

69. Ibid., 96.

70. Ibid., 107.

71. Ibid., 122.

72. Ibid.

73. Ibid., 101.

74. May, "Kneading," 6.
75. David Henningson, "Production Notes," *Somba-Ke: The Money Place*. http://www.sombake-themoneyplace.com/.
76. Ibid.
77. The World Nuclear Association suggests that talk of a "renaissance" began "around 2001," but they also argue that it only looked like a "renaissance" from the perspective of the "Western world," which had seen a lull previously. See "The Nuclear Renaissance" at http://www.world-nuclear.org/info/inf104.html.
78. David Henningson, *Somba-Ke: The Money Place* (Canada: Urgent Service Films, 2007), 55 min., video.
79. The CDUT Final Report Summary is available online at http://www.aadnc-aandc.gc.ca/eng/1100100023105/1100100023107.
80. Quoted in A. M. Muir, D. M. Leonard, and C. C. Krueger, "Past, Present and Future of Fishery Management on One of the World's Last Remaining Pristine Great Lakes: Great Bear Lake, Northwest Territories, Canada," *Reviews in Fish Biology and Fisheries* 23 (2013): 294.
81. Quoted in ibid., 295.
82. "Alberta Star Reports Assay Results from its 2008 Drilling Program" (November 24, 2008). http://www.alberta-star.com/s/NewsReleases.asp?ReportID=329252&_Type=News-Releases&_Title=Alberta-Star-Reports-Assay-Results-from-its-2008-Drilling-Program.
83. In her brilliant dissertation-in-progress, Alana Fletcher reports that in her interviews with the community in Deline, she found no support for uranium mining. See "Re/Mediation: The Story of Port Radium," PhD diss., Queen's University.
84. Gordon Pitts, "Exit Interview: Cameco CEO Spends Last Days Fighting Nuclear Panic," *Globe and Mail* (April 28, 2011). http://www.theglobeandmail.com/report-on-business/rob-magazine/cameco-ceo-spends-last-days-fighting-nuclear-panic/article624349/.
85. Susie O'Brien, "From TINA to TANIA: Shell Energy Scenarios and the Politics of Resilience," paper presented at the Association for Literature, Environment and Culture in Canada Conference, Thunder Bay, Ontario, August 7–10, 2014.
86. Pitts, "Exit Interview."
87. Klein, *This Changes Everything*, 58.
88. Quoted in David Lochbaum, Edwin Lyman, Susan Stranahan, and the Union of Concerned Scientists, *Fukushima: The Story of a Nuclear Disaster* (New York: New Press, 2014).
89. Lochbaum et al., *Fukushima*, 177.
90. Ibid.
91. Justin McCurry, "Doubts over Ice Wall to Keep Fukushima Safe from Damaged Nuclear Reactors," *Guardian* (July 13, 2014). http://www.theguardian.com/environment/2014/jul/13/doubts-giant-ice-wall-fukushima-nuclear-reactors. Given that this timeline comes from the plant's current manager, and TEPCO apologist, one assumes this is a highly optimistic projection.

92. McCurry, "Doubts."

93. Ibid.

94. Quoted in Vivien Diniz, "Uranium: The Short Term Is Bad, but the Long Term Looks Great," *Uranium Investing News* (July 2, 2014). http://uraniuminvestingnews.com/18684/uranium-the-short-term-is-bad-but-the-long-term-looks-great.html.

95. TEPCO asserted that they could not be held responsible for a natural disaster on the scale of the tsunami—which was "unimaginable" and "unpredictable" in nature. A subsequent report by an international panel of investigators, found, however, that "the government and TEPCO failed to prevent the crisis not because such a large tsunami was unanticipated but because they were reluctant to invest time, effort and money in protecting against a natural disaster considered unlikely." See Mari Yamaguchi, "Fukushima Nuclear Disaster Report: Plant Operators Tokyo Electric and Government Still Stumbling," *Huffington Post* (July 23, 2012). http://www.huffingtonpost.com/2012/07/23/fukushima-dai-ichi-nuclear-plant-operators_n_1694476.html.

96. Lochbaum et al., *Fukushima*, point, for example, to flooding, a natural disaster that threatens a number of U.S. reactors sited downstream from major dams.

97. Catriona Sandilands, "Acts of Nature: Literature, Excess and Environmental Politics," in *Critical Collaborations: Indigeneity, Diaspora, and Ecology in Canadian Literary Studies*, ed. Smaro Kamboureli and Christl Verduyn (Waterloo: Wilfrid Laurier University Press, 2014), 138–39.

98. Annie Smith, "Atomies of Desire: Directing *Burning Vision* in Northern Alberta," *Canadian Theatre Review* 144 (Fall 2010): 54.

99. Quoted in ibid., 58.

100. In an essay on teaching *Burning Vision* in the activist classroom, Allison Hargreaves dubs the play, borrowing from the stage directions that open its final movement, "a precise instrument for seeing"—in this case, for seeing "that which normally can't be seen" (52). See Allison Hargreaves, "'A Precise Instrument for Seeing': Remembrance in *Burning Vision* and the Activist Classroom," *Canadian Theatre Review* 147 (Summer 2011), 49–54. Though it is true that the play offers important insight, the language of "precision" is, I think, misleading. For another perspective on the play's production and politics, see May.

101. Smith, "Atomies of Desire," 59.

102. Chris Wood, "The Last Great Water Fight," *Walrus* (October 2010). http://thewalrus.ca/the-last-great-water-fight/.

103. Lorien Nesbitt, "Traditional Ecological Knowledge Summary Report: The Impacts of Climate Change on the Aquatic Ecosystems of Great Bear Lake and Its Watershed," report prepared for the Deline Renewable Resources Council by Lorien Environmental Consulting (January 28, 2011). Available at www.greatbearlake.org.

104. Clements, *Burning Vision*, 120.

105. Nick Mansfield, "The Future of the Future," in *Criticism, Crisis, and Contemporary Narrative: Textual Horizons in an Age of Risk*, ed. Paul Crosthwaite (New York: Routledge, 2011), 43.

106. Ulrich Beck, *Twenty Observations on a World in Turmoil*, trans. Ciaran Cronin (Cambridge: Polity Press, 2012), 107.
107. Van Wyck, *Highway of the Atom*, 49.
108. Beck, *World at Risk*, 35.
109. See Jonathan Watts, "Fukushima Parents Dish the Dirt in Protest over Radiation Levels," *Guardian* (May 2, 2011). http://www.guardian.co.uk/world/2011/may/02/parents-revolt-radiation-levels.
110. Ibid.
111. Lochbaum et al., *Fukushima*, 162.
112. Mansfield, "Future of the Future," 41.
113. Van Wyck, *Signs of Danger*.
114. U.S. Department of Energy, "What Happened at WIPP in February 2014." http://www.wipp.energy.gov/wipprecovery/accident_desc.html.
115. Beck, *Twenty Observations*, 106.
116. Williams, *Refuge*, 285.
117. Jacques Derrida, "Force of Law: The 'Mystical Foundation of Authority,'" in *Deconstruction and the Possibility of Justice*, ed. Drucilla Cornell, Michel Rosenfeld, and David Gray Carlson (New York: Routledge, 1992), 28.
118. Williams, *Refuge*, 290.
119. Cohen describes climate change as "cinematic" in "Anecographics," 42. His discussion of the material appears in "Introduction: Murmurations—'Climate Change' and the Defacement of Theory," in *Telemorphosis: Theory in the Era of Climate Change*, vol. 1, ed. Tom Cohen (Ann Arbor: Open Humanities Press, 2012), 31.
120. Van Wyck, *Highway of the Atom*, 83.

AFTERWORD

1. Qtd. in Rodolphe Gasché, "European Memories: Jan Patocka and Jacques Derrida on Responsibility," in *Derrida and the Time of the Political*, ed. Pheng Cheah and Suzanne Guerlac. (Durham: Duke University Press, 2009), 136.
2. Paul Voosen, "Geologists Drive Golden Spike toward Anthropocene's Base," *E&E Publishing* (September 17, 2012). http://www.eenews.net/stories/1059970036.
3. John Hendrix, "Doomsday," *Bulletin of the Atomic Scientists* (January–February 2007): 24–25.
4. Charlotte Eubanks, "The Mirror of Memory: Constructions of Hell in the Marukis' Murals," *PMLA* 124.5 (2009): 1614–31.
5. Iri Maruki, interview with John Junkerman, "Oil and Water: An Interview with the Artists," in *The Hiroshima Murals: The Art of Iri and Toshi Maruki*, ed. John W. Dower and John Junkerman (Tokyo: Kodansha International, 1985), 126.
6. Toshi Maruki, *Hiroshima No Pika* (New York: Lothrop, Lee & Shepard Books, 1982).
7. For more on the lives and works of the Marukis, see John Junkerman's film, *Hellfire* (1986) and John Dower and John Junkerman's coedited art book, *The Hiroshima Murals*.

8. Dower, "War, Peace, and Beauty: The Art of Iri Maruki and Toshi Maruki," in Dower and Junkerman, *Hiroshima Murals*, 10.

9. Quoted in "Oil and Water," 125.

10. Quoted in Dower and Junkerman, *Hiroshima Murals*, 127.

11. I take this translation from the website for the Maruki gallery, http://www.aya.or.jp/~marukimsn/gen/gen6e.html. Dower and Junkerman have a slightly different translation in their book, *Hiroshima Murals*.

12. Quoted in Richard Minear, "Review: The Atomic-Bomb Paintings," *Bulletin of Concerned Asian Scholars* 19.4 (1987): 60.

13. Eubanks, "Mirror of Memory," 1628.

14. Ibid.

15. Ibid.

16. James Heffernan, "Ekphrasis and Representation," *New Literary History* 22.2 (Spring 1991): 299.

17. Ibid., 301.

18. Eubanks, "Mirror of Memory," 1615.

19. Ronald Wallace, "Hell Mural: Panel I," in *The Makings of Happiness* (Pittsburgh: University of Pittsburgh Press, 1991), 35–36, © 1991 by Ronald Wallace, reprinted by permission of the University of Pittsburgh Press.

20. Heffernan, "Ekphrasis and Representation," 300.

21. My thanks to Peg Wallace for introducing me to this film, which was funded in part by the Wisconsin Humanities Council for which she worked.

22. Ronald Wallace, "Poems: 'Toads, and All This Fiddle: Wallace on Music, Metaphor, and Mirth,'" in *Contemporary American Poetry: Behind the Scenes*, ed. Ryan Van Cleave (New York: Longman, 2003), 294.

23. See W. J. T. Mitchell, "Ekphrasis and the Other," in *Picture Theory* (Chicago: University of Chicago Press, 1994).

24. Wallace, "The Hell Mural: Panel II," in *The Makings of Happiness*, 37.

Bibliography

Adams, Marty. "A Letter from Marty Adams, Director of Water Quality & Operations." Posted June 25, 2008. Silver Lake Reservoir Conservancy. http://www.silverlakereservoirs.org/news-and-events/bird-balls-on-ivanhoe-reservoir-are-non-toxic-questions-about-ivanhoe-bird-ball-safety-and-dwps-intentions/ (accessed February 24, 2015).
Agaral, Anil, Juliet Merrifield, and Rajessh Tandon. *No Place to Run: Local Realities and Global Issues of the Bhopal Disaster*. New Market, TN: Highlander Center; New Delhi: Society for Participatory Research in Asia, 1985.
Alaimo, Stacy. *Bodily Natures: Science, Environment, and the Material Self*. Bloomington: Indiana University Press, 2010.
Alaimo, Stacy. "States of Suspension: Trans-corporeality At Sea." *ISLE: Interdisciplinary Studies in Literature and Environment* 19.3 (Summer 2012): 476–93.
"Alberta Star Reports Assay Results from Its 2008 Drilling Program." November 24, 2008. http://www.alberta-star.com/s/NewsReleases.asp?ReportID=329252&_Type=News-Releases&_Title=Alberta-Star-Reports-Assay-Results-from-its-2008-Drilling-Program (accessed February 24, 2015).
Allison, Rachel Hope. *I'm Not a Plastic Bag: A Graphic Novel*. Los Angeles: Archaia, 2012.
Alworth, David. "Supermarket Sociology." *New Literary History* 41 (2010): 301–27.
Andersen, Ross. "After 4 Years, Checking Up on the Svalbard Global Seed Vault." *Atlantic* (February 28, 2012). http://www.theatlantic.com/technology/archive/2012/02/after-4-years-checking-up-on-the-svalbard-global-seed-vault/253458/ (accessed February 24, 2015).
Anderson, Jon. "50 Years after Hiroshima: Nuclear Protesters Collect Their Thoughts." *Chicago Tribune* (August 4, 1995). http://articles.chicagotribune.com/1995-08-04/news/9508040365_1_hiroshima-bomb-uranium-atom (accessed February 24, 2015).
Appadurai, Arjun. *Modernity at Large*. Minneapolis: University of Minnesota Press, 1996.
Bainbridge, Janet. "The Use of Substantial Equivalence in the Risk Assessment of GM Food." May 16, 2001. www.royalsociety.org (accessed September 27, 2009).
Bambara, Toni Cade. *The Salt Eaters*. New York: Vintage, 1981.
Bannet, Eve Tavor. "Analogy as Translation: Wittgenstein, Derrida, and the Law of Language." *New Literary History* 28 (1997): 655–72.

Barthes, Roland. *Mythologies*. Translated by Annette Lavers. New York: Hill and Wang, 1972.
Bauman, Zygmunt. *Liquid Modernity*. Malden, MA: Polity, 2000.
Beck, Ulrich. *Ecological Enlightenment: Essays on the Politics of the Risk Society*. Translated by Mark A. Ritter. Atlantic Highlands, NJ: Humanities Press, 1995.
Beck, Ulrich. "Interview/ Ulrich Beck: System of Organized Irresponsibility behind the Fukushima Crisis." Interview with Hirohito Ohno. *Asahi Shimbun* (July 6, 2011). http://ajw.asahi.com/article/0311disaster/analysis_opinion/AJ201107063167 (accessed February 24, 2015).
Beck, Ulrich. "Living in the World Risk Society." *Economy and Society* 35.3 (August 2006): 329–45.
Beck, Ulrich. *Risk Society*. Translated by Mark Ritter. London: Sage, 1992.
Beck, Ulrich. *Twenty Observations on a World in Turmoil*. Translated by Ciaran Cronin. Cambridge: Polity Press, 2012.
Beck, Ulrich. *World at Risk*. Translated by Ciaran Cronin. Cambridge: Polity Press, 2009.
Beck, Ulrich. *World Risk Society*. Translated by Ciaran Cronin. Cambridge: Polity Press, 1999.
Bennett, Jane. *Vibrant Matter: A Political Ecology of Things*. Durham: Duke University Press, 2010.
Bernstein, Barton. Introduction to *Voice of the Dolphins and Other Stories* by Leo Szilard, expanded ed., 3–43. Stanford: Stanford University Press, 1992.
Bérubé, Michael. "Plot Summary: Motives and Narrative Mechanics in *Underworld* and *White Noise*." In *Approaches to Teaching DeLillo's "White Noise,"* edited by Tim Engles and John N. Duvall, 135–43. New York: Modern Language Association, 2006.
Bilbrey, Jenna. "BPA-Free Plastic Containers May Be Just as Hazardous." *Scientific American* (August 11, 2014). http://www.scientificamerican.com/article/bpa-free-plastic-containers-may-be-just-as-hazardous/ (accessed February 24, 2015).
Black, Edward. "Top TV Scientist Backs GM Crops, but Fears Cloning." *Scotsman* (August 15, 2003). http://www.bic.searca.org/news/2003/aug/eur/15.html (accessed February 24, 2015).
Black, Richard. "Fukushima: As Bad as Chernobyl?" *BBC News* (April 12, 2011). http://www.bbc.co.uk/news/science-environment-13048916 (accessed February 24, 2015).
Blow, Peter. *Village of Widows*. Toronto: Lindum Films, 1998. Video.
Board of Directors. "It Is Five Minutes to Midnight." *Bulletin of the Atomic Scientists* (January–February 2007): 66–71.
Board of Directors. "It Is Six Minutes to Midnight." Press Release. *Bulletin of the Atomic Scientists* (January 14, 2010). http://turnbacktheclock.org/press-release/it-6-minutes-midnight (accessed July 28, 2015).
"Body 2.0: Creating a World That Can Feed Itself." Zeitgeist: The Google Partner Forum. YouTube video. 36:18. Uploaded by zeitgeist08. September 18, 2008. https://www.youtube.com/watch?v=I9I1IkbcHNE (accessed September 8, 2015).

Boyer, Paul. *By the Bomb's Early Light*. Chapel Hill: University of North Carolina Press, 1985.
Bradford, David. "We All Live in Bhopal." In *The Bhopal Reader*, edited by Bridget Hanna, Ward Morehouse, and Satinath Sarangi, 283–86. New York: Apex Press, 2005.
Bradford, David. "We All Live in Bhopal." *Fifth Estate* (Winter 1985). http://www.eco-action.org/dt/bhopal.html (accessed February 24, 2015).
Bradley, David. *No Place to Hide*. Boston: Little, Brown, 1948.
Brians, Paul. "Farewell to the First Atomic Age." *Nuclear Texts and Contexts* 8 (Fall 1992): 1–3.
Brown, Bill. "How to Do Things with Things (a Toy Story)." *Critical Inquiry* 24.4 (Summer 1998): 935–64.
Buell, Frederick. *From Apocalypse to Way of Life: Environmental Crisis in the American Century*. New York: Routledge, 2004.
Buell, Lawrence. Foreword to *Environmental Criticism for the Twenty-First Century*, edited by Stephanie LeMenager, Teresa Shewry, and Ken Hiltner, xiii–xvii. New York: Routledge, 2011.
Buell, Lawrence. *The Future of Environmental Criticism: Environmental Crisis and Literary Imagination*. Malden, MA: Blackwell, 2005.
Buell, Lawrence. "Toxic Discourse." *Critical Inquiry* 24.3 (Spring 1998): 639–65.
Buell, Lawrence. *Writing for an Endangered World*. Cambridge: Harvard University Press, 2001.
Burke, Kenneth. "Four Master Tropes." *Kenyon Review* 3.4 (Autumn 1941): 421–38.
Carrigan, Anthony. "'Justice Is on Our Side'? Animal's People, Generic Hybridity, and Eco-crime." *Journal of Commonwealth Literature* 47.2 (2012): 159–74.
Carruth, Allison. "The City Refigured: Environmental Vision in a Transgenic Age." In *Environmental Criticism for the Twenty-First Century*, edited by Stephanie LeMenager, Teresa Shewry, and Ken Hiltner, 85–104. London: Routledge, 2011.
Carruth, Allison. *Global Appetites*. Cambridge: Cambridge University Press, 2013.
Carson, Rachel. *Silent Spring*. Boston: Houghton Mifflin, 1962.
Caruth, Cathy. "Afterword: Turning Back to Literature." *PMLA* 125.4 (2010): 1087–95.
Chakrabarty, Dipesh. "The Climate of History: Four Theses." *Critical Inquiry* 35 (Winter 2009): 197–222.
Chaloupka, William. *Knowing Nukes: The Politics and Culture of the Atom*. Minneapolis: University of Minnesota Press, 1992.
Charles, Ron. "Starch-Free Protest." Review of *All Over Creation*, by Ruth Ozeki. *Christian Science Monitor* (March 13, 2003). http://www.ruthozeki.com/reviews/csm.html (accessed September 28, 2009).
Chávez, César. *The Words of César Chávez*. Edited by Richard Jensen and John Hammerback. College Station: Texas A&M University Press, 2002.
Chávez, César. "The Wrath of Grapes." In *The Words of César Chávez*, edited by Richard Jensen and John Hammerback, 132–35. College Station: Texas A&M University Press, 2002.

Chiu, Monica. "Postnational Globalization and (En)Gendered Meat Production in Ruth L. Ozeki's *My Year of Meats.*" *LIT: Literature, Interpretation, Theory* 12.1 (2001): 99–128.

Cixous, Hélène. "Laugh of the Medusa." *Signs* 1.4 (Summer 1976): 875–93.

Clapp, Jennifer. "The Political Economy of Food Aid in an Era of Agricultural Biotechnology." *Global Governance* 11 (2005): 467–85.

Clark, Timothy. "Toward a Deconstructive Environmental Criticism." *Oxford Literary Review* 30.1 (2008): 45–68.

Clayton, Jay. "Victorian Chimeras, or, What Literature Can Contribute to Genetics Policy Today." *New Literary History* 38 (2007): 569–91.

Clements, Marie. *Burning Vision.* Vancouver: Talon Books, 2003.

Cohen, Judith Beth. "Bad Seeds." Review of *All Over Creation,* by Ruth Ozeki. *Women's Review of Books* 20.8 (May 2003): 6.

Cohen, Tom. "Anecographics." In *Impasses of the Post-global: Theory in an Age of Climate Change,* edited by Henry Sussman, 32–57. Ann Arbor: Open Humanities Press, 2012.

Cohen, Tom. "Introduction: Murmurations—'Climate Change' and the Defacement of Theory." In *Telemorphosis: Theory in the Era of Climate Change,* edited by Tom Cohen, 13–42. Ann Arbor: Open Humanities Press, 2012.

Collins, Janelle. "Generating Power: Fission, Fusion, and Postmodern Politics in Bambara's *The Salt Eaters.*" *MELUS: Multi-ethnic Literature of the United States* 21.2 (Summer 1996): 35–47.

Conley, Verena Andermatt. *Ecopolitics: The Environment in Poststructuralist Thought.* New York: Routledge, 1997.

Cooper, Ken. "The Whiteness of the Bomb." In *Postmodern Apocalypse: Theory and Practice at the End,* edited by Richard Dellamora, 79–106. Philadelphia: University of Pennsylvania Press, 1995.

Cordle, Daniel. "Cultures of Terror: Nuclear Criticism during and since the Cold War." *Literature Compass* 3.6 (2006): 1186–99.

Cordle, Daniel. *States of Suspense: The Nuclear Age, Postmodernism and United States Fiction and Prose.* Manchester: Manchester University Press, 2008.

Cummings, Claire Hope. *Uncertain Peril: Genetic Engineering and the Future of Seeds.* Boston: Beacon Press, 2008.

Davies, D. R. "Atomic Energy and Personal Responsibility." Foreword to *One World or None,* edited by Dexter Masters and Katherine Way. London: Latimer House Limited, 1947.

Davis, Mike. *Planet of Slums.* London: Verso, 2007.

de Man, Paul. *Allegories of Reading.* New Haven: Yale University Press, 1979.

de Man, Paul. "Anthropomorphism and Trope in Lyric." In *The Rhetoric of Romanticism,* 239–62. New York: Columbia University Press, 1984.

de Turenne, Veronique. "Those Reservoir Balls—Are They Safe?" *Los Angeles Times* (June 10, 2008). http://latimesblogs.latimes.com/lanow/2008/06/franciscos-post.html (accessed February 26, 2015).

DeLillo, Don. *Underworld.* New York: Scribner, 1997.

DeLillo, Don. *White Noise.* Edited by Mark Osteen. New York: Penguin, 1998.

DeLoughrey, Elizabeth. "Radiation Ecologies and the Wars of Light." *Modern Fiction Studies* 55.3 (Fall 2009): 468–98.

Deresiewicz, William. "'I Was There': On Kurt Vonnegut." *Nation* (June 4, 2012). http://www.thenation.com/article/167921/i-was-there-kurt-vonnegut #axzz2bmPyvjpO (accessed February 24, 2015).
Derrida, Jacques. *Archive Fever*. Translated by Eric Prenowitz. Chicago: University of Chicago Press, 1998.
Derrida, Jacques. "Autoimmunity: Real and Symbolic Suicides—a Dialogue with Jacques Derrida." In *Philosophy in a Time of Terror: Dialogues with Jürgen Habermas and Jacques Derrida*, edited by Giovanna Borradori, 85–136. Chicago: University of Chicago Press, 2003.
Derrida, Jacques. "Biodegradables: Seven Diary Fragments." Translated by Peggy Kamuf. *Critical Inquiry* 15.4 (Summer 1989): 812–73.
Derrida, Jacques. "Force of Law: The 'Mystical Foundation of Authority.'" In *Deconstruction and the Possibility of Justice*, edited by Drucilla Cornell, Michel Rosenfeld, and David Gray Carlson, 3–67. New York: Routledge, 1992.
Derrida, Jacques. "No Apocalypse, Not Now." Translated by Catherine Porter and Philip Lewis. *Diacritics* 14 (Summer 1984): 20–31.
Dickinson, Adam. *The Polymers*. Toronto: House of Anansi Press, 2013.
Diniz, Vivien. "Uranium: The Short Term Is Bad, but the Long Term Looks Great." *Uranium Investing News* (July 2, 2014). http://uraniuminvesting-news.com/18684/uranium-the-short-term-is-bad-but-the-long-term-looks-great.html (accessed February 24, 2015).
"'Doomsday Clock' Moves Two Minutes Closer to Midnight." Press Release of the *Bulletin of the Atomic Scientists*, January 17, 2007. http://thebulletin.org/press-release/doomsday-clock-moves-two-minutes-closer-midnight (accessed February 24, 2015).
Dower, John, and John Junkerman, eds. *The Hiroshima Murals: The Art of Iri and Toshi Maruki*. Tokyo: Kodansha International, 1985.
Duvall, John. "The (Super)Market of Images: Television as Unmediated Mediation in DeLillo's *White Noise*." In *White Noise: Text and Criticism*, edited by Mark Osteen, 432–55. New York: Penguin, 1998.
Dyson, Freeman. "Our Biotech Future." In *The Best American Science and Nature Writing*, edited by Jerome Groopman, 64–74. Boston: Houghton Mifflin, 2008.
Ehrlich, Paul. *The Population Bomb*. New York: Ballantine, 1968.
Eubanks, Charlotte. "The Mirror of Memory: Constructions of Hell in the Marukis' Murals." *PMLA* 124.5 (2009): 1614–31.
Evans, Mei Mei, and Rachel Stein, eds. *The Environmental Justice Reader*. Tucson: University of Arizona Press, 2002.
Ewald, François. "The Return of Descartes's Malicious Demon: An Outline of a Theory of Precaution." In *Embracing Risk*, edited by Tom Baker and Jonathan Simon, 273–301. Chicago: University of Chicago Press, 2002.
Fackler, Martin. "Leak Found in Steel Tank for Water at Fukushima." *New York Times* (June 5, 2013). http://www.nytimes.com/2013/06/06/world/asia/tepco-says-water-at-fukushima-is-contaminated.html?_r=2& (accessed February 25, 2015).
Ferguson, Frances. "The Nuclear Sublime." *Diacritics* 14.2 (Summer 1984): 4–10.
Fermi, Enrico. "The Development of the First Chain Reacting Pile." *Proceedings of the American Philosophical Society* 90.1 (January 1946): 20–24.

Fermi, Enrico. "Fermi's Own Story." *Chicago Sun-Times* (November 23, 1952). Included in the U.S. Department of Energy's Report, *The First Reactor* (December 1982): 21–26. http://www.osti.gov/accomplishments/documents/fullText/ACC0044.pdf (accessed February 25, 2015).

Fletcher, Alana. "Re/mediation: The Story of Port Radium." PhD diss., Queen's University, 2015.

Fortun, Kim. *Advocacy after Bhopal: Environmentalism, Disaster, New Global Orders*. Chicago: University of Chicago Press, 2001.

Foucault, Michel. *Power/Knowledge: Selected Interviews and Other Writings, 1972–1977*. Edited by Colin Gordon. Translated by Colin Gordon, Leo Marshall, John Mepham, and Kate Soper. New York: Pantheon, 1972.

Foucault, Michel. *"Society Must Be Defended."* London: Picador 2003.

François, Anne-Lise. "'Oh Happy Living Things': Frankenfoods and the Bounds of Wordsworthian Natural Piety." *Diacritics* 33.2 (Summer 2003): 42–70.

Freinkel, Susan. *Plastic: A Toxic Love Story*. New York: Houghton Mifflin Harcourt, 2011.

Garrard, Greg. "Nature Cures? or How to Police Analogies of Personal and Ecological Health." *ISLE: Interdisciplinary Studies in Literature and Environment* 19.3 (Summer 2012): 494–514.

Gasché, Rodolphe. "European Memories: Jan Patocka and Jacques Derrida on Responsibility." In *Derrida and the Time of the Political*, edited by Pheng Cheah and Suzanne Guerlac, 135–57. Durham: Duke University Press, 2009.

Gayle, Damien. "Could a Plastic-Eating Fungi Save the World from Biggest Man-Made Environmental Catastrophe?" *Daily Mail* (May 18, 2012). http://www.dailymail.co.uk/sciencetech/article-2146224/Could-fungi-break-plastic-stop-modern-scourge.html (accessed February 25, 2015).

"Genetically Modified Crops Get Vatican's Blessing." *New Scientist* (June 4, 2009). http://www.newscientist.com/article/mg20227114.200-genetically-modified-crops-get-the-vaticans-blessing.html (accessed August 15, 2009).

Godmilow, Jill. Interview with Ann-Louise Shapiro. "How Real Is the Reality in Documentary Film?" *History and Theory* 36.4 (December 1997): 80–101.

GRAIN. "Faults in the Vault: Not Everyone Is Celebrating Svalbard" (February 2008). http://www.grain.org/fr/article/entries/181-faults-in-the-vault-not-everyone-is-celebrating-svalbard (accessed February 25, 2015).

Grausam, Daniel. "'It Is Only a Statement of the Power of What Comes After': Atomic Nostalgia and the Ends of Postmodernism." *American Literary History* 24.2 (2012): 308–36.

Guha, Ramachandra, and Juan Martinez-Alvier. *Varieties of Environmentalism: Essays North and South*. Oxford: Oxford University Press, 1997.

Hacking, Ian. "Our Neo-Cartesian Bodies in Parts." *Critical Inquiry* 34 (Autumn 2007): 78–105.

Hamblin, Jacob Darwin. "Fukushima and the Motifs of Nuclear History." *Environmental History* 17 (April 2012): 285–99.

Handwerk, Brian. "Doomsday Seed Vault Safeguards Our Food Supply." *National Geographic* (July 2, 2012). http://news.nationalgeographic.com/news/pictures/2012/07/120702-svalbard-doomsday-seed-vault-food-supply/ (accessed February 25, 2015).

Haraway, Donna. *The Companion Species Manifesto*. Chicago: Prickly Paradigm Press, 2003.
Haraway, Donna. *Modest_Witness@Second_Millennium*. New York: Routledge, 1997.
Haraway, Donna. *Simians, Cyborgs, and Women*. New York: Routledge, 1991.
Hardt, Michael, and Antonio Negri. *Multitude*. New York: Penguin, 2004.
Hargreaves, Allison. "'A Precise Instrument for Seeing': Remembrance in *Burning Vision* and the Activist Classroom." *Canadian Theatre Review* 147 (Summer 2011): 49–54.
Hawkins, Gay. *The Ethics of Waste: How We Relate to Rubbish*. New York: Rowman & Littlefield, 2005.
Hawkins, Gay. "Made to Be Wasted: PET and the Topologies of Disposability." In *Accumulation: The Material Politics of Plastic*, edited by Jennifer Gabrys, Gay Hawkins, and Mike Michael, 49–67. London: Routledge, 2014.
Hawkins, Gay. "Plastic Materialities." In *Political Matter: Technoscience, Democracy, and Public Life*. Edited by Bruce Braun and Sarah J. Whatmore, 119–38. Minneapolis: University of Minnesota Press, 2010.
"HDPE Pipe for Potable Water Applications." Report by the PPRC. February 2010. http://www.google.com/url?sa=t&rct=j&q=&esrc=s&source=web&cd=9&ved=0CGcQFjAI&url=http%3A%2F%2Fwww.pprc.org%2Fresearch%2FrapidresDocs%2FPPRC_HDPE_Water_Pipe_Safety_FINAL.pdf&ei=z9CxUsWcHsTgyQG91oGoAw&usg=AFQjCNEwwrFvujtQwT0FXB1rgqCu3wLQKg&bvm=bv.58187178,d.aWc (accessed February 25, 2015).
Heal the Bay. *The Majestic Plastic Bag*. Video. Directed by Jeremy Konner. 2010. Santa Monica. http://www.healthebay.org/media-center/videos (accessed February 25, 2015).
Heffernan, James. "Ekphrasis and Representation." *New Literary History* 22.2 (Spring 1991): 297–316.
Heise, Ursula. "Ecocriticism and the Transnational Turn in American Studies." *American Literary History* 20.1–2 (Spring–Summer 2008): 1–24.
Heise, Ursula. *Sense of Place and Sense of Planet*. New York: Oxford University Press, 2008.
Hendrix, John. "Doomsday." *Bulletin of the Atomic Scientists* (January–February 2007), 24–25.
Henningson, David. *Somba-Ke: The Money Place*. Canada: Urgent Service Films, 2007. Video.
Herring, Ronald. "Stealth Seeds: Bioproperty, Biosafety, Biopolitics." *Journal of Development Studies* 43.1 (January 2007): 130–57.
Herring, Ronald. "Why Did 'Operation Cremate Monsanto' Fail? Science and Class in India's Great Terminator-Technology Hoax." *Critical Asian Studies* 38.4 (2006): 467–93.
Hird, Myra. "Knowing Waste: Toward an Inhuman Epistemology." *Social Epistemology* 26.3–4 (2012): 453–69.
Iovino, Serenella, and Serpil Oppermann. "Theorizing Material Ecocriticism: A Diptych." *ISLE: Interdisciplinary Studies in Literature and Environment* 19.3 (Summer 2012): 448–75.
Jakobson, Roman. "Two Aspects of Language and Two Types of Aphasic Dis-

turbances." In *On Language,* edited by Linda R. Waugh and Monique Monville Burston, 115–33. Cambridge: Harvard University Press, 1971.

"Japan Reactors at 'High Risk' for More Leaks." *CBC News* (March 15, 2011). http://news.aol.ca/2011/03/15/japan-reactors-at-high-risk-for-more-leaks-pm/ (accessed April 30, 2012).

Jasanoff, Sheila. "Bhopals Trials of Knowledge and Ignorance." *ISIS* 98.2 (2007): 344–50.

Jasanoff, Sheila. "Risk in Hindsight: Toward a Politics of Reflection." In *Risk Society and the Culture of Precaution,* edited by Ingo Richter, Sabine Berking, and Ralf Müller-Schmidt, 28–46. New York: Palgrave Macmillan, 2006.

Jinnah, Sikina. "Climate Change Bandwagoning: The Impacts of Strategic Linkages on Regime Design, Maintenance, and Death." *Global Environmental Politics* 11.3 (August 2011): 1–9.

Johnson, Barbara. *Persons and Things.* Cambridge: Harvard University Press, 2010.

Junkerman, John. *Hellfire.* DVD. Directed by John Junkerman. 1986. Knightscove-Ellis International, 2005.

Kaniewski, Wojciech K., and Peter E. Thomas. "The Potato Story." *AgBioForum* 7.1–2 (2004): 41–46.

Kasperson, Jeanne X., Roger E. Kasperson, Nick Pidgeon, and Paul Slovic. "The Social Amplification of Risk: Assessing Fifteen Years of Research and Theory." In *The Social Amplification of Risk,* ed. Nick Pidgeon, Roger Kasperson, and Paul Slovic, 13–47. Cambridge: Cambridge University Press, 2003.

Kass, Leon, ed. *Being Human: Core Readings in the Humanities.* New York: Norton, 2004.

Keen, Sam. "A Secular Apocalypse." *Bulletin of the Atomic Scientists* (January 2007): 29–31.

Kermode, Frank. *The Sense of an Ending.* 2nd ed. New York: Oxford University Press, 2000.

Klein, Naomi. *This Changes Everything: Capitalism vs. the Climate.* Toronto: Knopf Canada, 2014.

Klein, Richard. "The Future of Nuclear Criticism." *Yale French Studies* 77 (1990): 76–100.

Kohso, Sabu. "Rise of the New Collective Intellect—from Apocalyptic Disaster and Mass Insurrection (2011)." *Inter-Asia Cultural Studies* 13.1 (2012): 159–64.

Kuiper, Harry A., Gijs A. Kleter, Hub P. J. M. Noteborn, and Esther J. Kok. "Substantial Equivalence—an Appropriate Paradigm for the Safety Assessment of Genetically Modified Foods?" *Toxicology* 181–182 (2002): 427–31.

Ladino, Jennifer. *Reclaiming Nostalgia: Longing for Nature in American Literature.* Charlottesville: University of Virginia Press, 2012.

Ladino, Jennifer. "Unlikely Alliances: Notes on a Green Culture of Life." *Journal of Religion and Society* (2008): 146–58.

Lapierre, Dominique, and Javier Moro. *Five Past Midnight in Bhopal.* New York: Warner Books, 2002.

Lamb, Jonathan. *The Things Things Say.* Princeton: Princeton University Press, 2011.

Latour, Bruno. "Morality and Technology: The End of the Means." Translated by Couze Venn. *Theory, Culture and Society* 15.5–6 (2002): 247–60.
Latour, Bruno. *Politics of Nature: How to Bring the Sciences into Democracy.* Translated by Catherine Porter. Cambridge: Harvard University Press, 2004.
Latour, Bruno. "Why Has Critique Run Out of Steam? From Matters of Fact to Matters of Concern." *Critical Inquiry* 30.2 (Winter 2004): 225–48.
LeGuin, Ursula. *The Left Hand of Darkness.* New York: Ace, 1976.
Leiss, William, and Douglas Powell. *Mad Cows and Mother's Milk: The Perils of Poor Risk Communication.* 2nd ed. Montreal: McGill-Queen's University Press, 2004.
"Leo Seren, 83; Physicist on the Atomic Bomb Who Turned Pacifist." Obituaries, *Los Angeles Times* (January 11, 2002). http://articles.latimes.com/2002/jan/11/local/me-passings11.2 (accessed February 25, 2015).
Leslie, Keith. "Ontario Reactors Safe, Robust: Agency Review." *Globe and Mail* (August 23, 2012). http://m.theglobeandmail.com/technology/science/ontario-reactors-safe-robust-agency-review/article597971/?service=mobile (accessed February 25, 2015).
Liboiron, Max. "Plasticizers: A Twenty-First Century Miasma." In *Accumulation: The Material Politics of Plastic*, edited by Jennifer Gabrys, Gay Hawkins, and Mike Michael, 134–149. London: Routledge, 2014.
Lochbaum, David, Edwin Lyman, Susan Stranahan, and the Union of Concerned Scientists. *Fukushima: The Story of a Nuclear Disaster.* New York: New Press, 2014.
Lodge, David. *The Modes of Modern Writing: Metaphor, Metonymy and the Typology of Modern Literature.* London: Edward Arnold, 1977.
Long, Aaron. "Archaia Teams With JeffCorwinConnect for I'M NOT A PLASTIC BAG." Comicuriosity (February 23, 2012). http://www.comicosity.com/archaia-teams-with-jeffcorwinconnect-for-i-am-not-a-plastic-bag/ (accessed February 25, 2015).
Luckhurst, Roger. "Nuclear Criticism: Anachronism and Anachorism." *Diacritics* 23.2 (Summer 1993): 89–97.
Mahapatra, Dhananjay. "Threat of Plastic Bags Bigger Than the Atom Bomb: SC." *Times of India* (May 7, 2012). http://timesofindia.indiatimes.com/home/environment/pollution/Threat-of-plastic-bags-bigger-than-atom-bomb-SC/articleshow/13031361.cms (accessed February 25, 2015).
Mansfield, Nick. "The Future of the Future." In *Criticism, Crisis, and Contemporary Narrative: Textual Horizons in an Age of Risk*, edited by Paul Crosthwaite, 31–45. New York: Routledge, 2011.
Martucci, Elise. *The Environmental Unconscious in the Fiction of Don DeLillo.* New York: Routledge, 2007.
Maruki, Iri. "Oil and Water: An Interview with the Artists." Interview with John Junkerman. In *The Hiroshima Murals: The Art of Iri and Toshi Maruki*, edited by John W. Dower and John Junkerman, 121–28. Tokyo: Kodansha International, 1985.
Maruki, Toshi. *Hiroshima No Pika.* New York: Lothrop, Lee & Shepard Books, 1982.
May, Theresa. "Kneading Marie Clements' Burning Vision." *Canadian Theatre Review* 144 (Fall 2010): 5–12.

May, Theresa. "This Is My Neighborhood." Keynote address at "Staging Sustainability" conference, York University, Toronto, Ontario, April 20, 2011.

McCurry, Justin. "Doubts over Ice Wall to Keep Fukushima Safe from Damaged Nuclear Reactors." *Guardian* (July 13, 2014). http://www.theguardian.com/environment/2014/jul/13/doubts-giant-ice-wall-fukushima-nuclear-reactors (accessed February 25, 2015).

McHugh, Josh. "Google versus Evil." *Wired* 11.01. http://archive.wired.com/wired/archive/11.01/google_pr.html (accessed February 25, 2015).

McHugh, Susan. "Flora, Not Fauna: GM Culture and Agriculture." *Literature and Medicine* 26.1 (Spring 2007): 25–54.

McKibben, Bill. "Actions Speak Louder Than Words." Interview. *Bulletin of the Atomic Scientists* 68.2 (2012): 1–8.

McKibben, Bill. *Eaarth: Making a Life on a Tough New Planet*. Toronto: Knopf Canada, 2010.

McKibben, Bill. *The End of Nature*. New York: Random House, 1989.

McMurray, Andrew. "The Slow Apocalypse: A Gradualistic Theory of the World's Demise." *Postmodern Culture* 6.3 (1996). http://muse.jhu.edu/login?auth=0&type=summary&url=/journals/postmodern_culture/v006/6.3mcmurry.html (accessed July 28, 2015).

Messmer, Michael. "Thinking It Through Completely': The Interpretation of Nuclear Culture." *Centennial Review* 32.4 (Fall 1988): 397–413.

Mian, Zia. "Out of the Nuclear Shadow: Scientists and the Struggle against the Bomb." *Bulletin of the Atomic Scientists* 71.1 (2015): 59–69.

Miller, Henry I., Gregory Conko, and Drew L. Kershen. "Why Spurning Food Biotech Has Become a Liability." *Nature Biotechnology* 24.9 (September 2006): 1075–77.

Millet, Lydia. *Oh Pure and Radiant Heart*. New York: Soft Skull Press, 2005.

Millstone, Erik, Eric Brunner, and Sue Mayer. "Beyond 'Substantial Equivalence.'" *Nature* 401 (October 7, 1999): 525–26.

Minear, Richard. "Review: The Atomic-Bomb Paintings." *Bulletin of Concerned Asian Scholars* 19.4 (1987): 58–63.

Mitchell, W. J. T. *Picture Theory*. Chicago: University of Chicago Press, 1994.

Mizruchi, Susan. "Risk Theory and the Contemporary American Novel." *American Literary History* 22.1 (Spring 2010): 109–35.

Monbiot, George. "Why Fukushima Made Me Stop Worrying and Love Nuclear Power." *Guardian* (March 21, 2011). http://www.theguardian.com/commentisfree/2011/mar/21/pro-nuclear-japan-fukushima?commentpage=1 (accessed February 25, 2015).

Moore, Ward. *Greener Than You Think*. Rockville, MD: Wildside Press. Originally published 1947.

Morton, Timothy. *The Ecological Thought*. Cambridge: Harvard University Press, 2010.

Morton, Timothy. *Ecology without Nature: Rethinking Environmental Aesthetics*. Cambridge: Harvard University Press, 2007.

Morton, Timothy. *Hyperobjects: Philosophy and Ecology after the End of the World*. Minneapolis: University of Minnesota Press, 2013.

Morton, Timothy. "An Object-Oriented Defense of Poetry." *New Literary History* 43.2 (Spring 2012): 205–24.
Muecke, D. C. *The Compass of Irony*. London: Methuen, 1969.
Muir, A. M., D. M. Leonard, and C. C. Krueger. "Past, Present and Future of Fishery Management on One of the World's Last Remaining Pristine Great Lakes: Great Bear Lake, Northwest Territories, Canada." *Reviews in Fish Biology and Fisheries* 23 (2013): 293–315.
Nesbitt, Lorien. "Traditional Ecological Knowledge Summary Report: The Impacts of Climate Change on the Aquatic Ecosystems of Great Bear Lake and Its Watershed." Report prepared for the Deline Renewable Resources Council by Lorien Environmental Consulting (January 28, 2011). www.greatbearlake.org (accessed February 25, 2015).
Nixon, Rob. *Slow Violence and the Environmentalism of the Poor*. Cambridge: Harvard University Press, 2011.
National Oceanic and Atmospheric Association. "How Big Is the 'Great Pacific Garbage Patch'? Science vs. Myth." http://response.restoration.noaa.gov/about/media/how-big-great-pacific-garbage-patch-science-vs-myth.html (accessed July 28, 2015).
Norris, Christopher. "'Nuclear Criticism' Ten Years On." *Prose Studies* 17.2 (August 1994): 130–60.
Norris, Christopher. *Uncritical Theory: Postmodernism, Intellectuals, and the Gulf War*. London: Lawrence & Wishart, 1992.
Norris, Margot. "Dividing the Indivisible: The Fissured Story of the Manhattan Project." *Cultural Critique* 35 (Winter 1996–1997): 5–38.
"Nuclear Fears Prompt Rush for Pills in B.C." *CBC News* (March 14, 2011). http://www.cbc.ca/news/canada/british-columbia/story/2011/03/14/bc-japan-radiation-iodine-pills.html (accessed February 25, 2015).
O'Brien, Susie. "From TINA to TANIA: Shell Energy Scenarios and the Politics of Resilience." Paper presented at the Association for Literature, Environment and Culture in Canada conference, Thunder Bay, Ontario, August 7–10, 2014.
O'Brien, Susie. "Postcolonial Fiction and the Temporality of Global Risk." Paper presented at the "Time and Globalization" conference, the Workers Arts and Heritage Center, McMaster University, Hamilton, Ontario, October 19–20, 2012.
Oppenheimer, J. Robert. "Atomic Weapons." *Proceedings of the American Philosophical Society* (January 1946): 7–10.
Oppenheimer, J. Robert. *The Open Mind*. New York: Simon and Schuster, 1955.
Ozeki, Ruth. *All Over Creation*. New York: Viking, 2003.
Ozeki, Ruth. *My Year of Meats*. New York: Penguin, 1998.
Ozeki, Ruth. "Seeds of Our Stories." *Moebius* 6.1 (2008): 13–25.
Packer, Matthew. "'At the Dead Center of Things' in Don DeLillo's *White Noise*: Mimesis, Violence, and Religious Awe." *Modern Fiction Studies* 51.3 (Fall 2005): 648–66.
Palmer, James. "A Review of 'Plastic Bag.'" *Psychological Perspectives* 54.2 (2011): 243–46.

Palumbo-Liu, David. "Rational and Irrational Choices: Form, Affect, and Ethics." In *Minor Transnationalism*, edited by Francoise Lionnet and Shu-mei Shih, 41–72. Durham: Duke University Press, 2005.

Perrow, Charles. "Fukushima and the Inevitability of Accidents." *Bulletin of the Atomic Scientists* 67.6 (2011): 44–52.

Perrow, Charles. "Nuclear Denial: From Hiroshima to Fukushima." *Bulletin of the Atomic Scientists* 69.5 (2013): 56–67.

Phillips, Jayne Anne. "Crowding Out Death." *New York Times* (January 13, 1985). http://www.nytimes.com/books/97/03/16/lifetimes/del-r-white-noise.html (accessed February 25, 2015).

Pitts, Gordon. "Exit Interview: Cameco CEO Spends Last Days Fighting Nuclear Panic." *Globe and Mail* (April 28, 2011). http://www.theglobeandmail.com/report-on-business/rob-magazine/cameco-ceo-spends-last-days-fighting-nuclear-panic/article624349 / (accessed February 25, 2015).

Pollan, Michael. "Playing God in the Garden." *New York Times Magazine* (October 25, 1998). http://www.nytimes.com/1998/10/25/magazine/playing-god-in-the-garden.html (accessed October 1, 2009).

Potrykus, Ingo. "Golden Rice and Beyond." *Plant Physiology* 125 (March 2001): 1157–61.

Rabinowitch, Eugene. "Five Years After." *Bulletin of the Atomic Scientists* (January 1951): 3–12.

Rhodes, Richard. *Making the Atomic Bomb*. New York: Simon and Schuster, 1987.

Roberts, Jody. "Reflections of an Unrepentant Plastiphobe." In *Accumulation: The Material Politics of Plastic*, edited by Jennifer Gabrys, Gay Hawkins, and Mike Michael, 121–33. London: Routledge, 2014.

Robin, Marie-Monique. *World According to Monsanto*. English ed. Yes! Books, 2008. DVD.

Rochman, Chelsea M., Mark Anthony Browne, Benjamin S. Halpern, Brian T. Hentschel, Eunha Oh, Hrissi K. Karapanagioti, Lorena M. Rios-Mendoza, Hideshige Takada, Swee The, and Richard C. Thompson. "Classify Plastic Waste as Hazardous." *Nature* 494 (February 14, 2013): 169–71.

Rody, Caroline. "Impossible Voices: Ethnic Postmodern Narration in Toni Morrison's *Jazz* and Karen Tei Yamashita's *Through the Arc of the Rain Forest*." *Contemporary Literature* 41.4 (Winter 2000): 618–41.

Ruthven, Kenneth Knowles. *Nuclear Criticism*. Melbourne: Melbourne University Press, 1993.

Sandilands, Catriona. "Acts of Nature: Literature, Excess and Environmental Politics." In *Critical Collaborations: Indigeneity, Diaspora, and Ecology in Canadian Literary Studies*, edited by Smaro Kamboureli and Christl Verduyn, 127–42. Waterloo: Wilfrid Laurier University Press, 2014.

Schell, Jonathan. *The Fate of the Earth and the Abolition*. Stanford: Stanford University Press, 2000.

Schwenger, Peter. *Letter Bomb: Nuclear Holocaust and the Exploding Word*. Baltimore: Johns Hopkins University Press, 1992.

Shiva, Vandana. *Stolen Harvest: The Hijacking of the Global Food Supply*. Cambridge, MA: South End Press, 2000.

Sibree, Bron. "The Pen versus Poison." *Courier Mail* (July 7, 2007). http://www.bhopal.net/the-pen-versus-poison/ (accessed February 26, 2015).

Sinha, Indra. *Animal's People*. London: Simon and Schuster, 2007.
Sinha, Indra. "Bhopal: A Novel Quest for Justice." *Guardian* (October 10, 2007). http://www.theguardian.com/world/2007/oct/10/india-bhopal (accessed February 26, 2015).
Slovic, Paul. "The Perception Gap: Radiation and Risk." *Bulletin of the Atomic Scientists* 68.3 (2012): 67–75.
Smith, Alice Kimball. *A Peril and a Hope: The Scientists' Movement in America, 1945–47*. Chicago: University of Chicago Press, 1965.
Smith, Annie. "Atomies of Desire: Directing Burning Vision in Northern Alberta." *Canadian Theatre Review* 144 (Fall 2010): 54–59.
Smith, Jeffrey. *Seeds of Deception: Exposing Industry and Government Lies about the Safety of the Genetically Engineered Foods You're Eating*. Fairfield, IA: Yes! Books, 2003.
Snell, Heather. "Assessing the Limitations of Laughter in Indra Sinha's *Animal's People*." *Postcolonial Text* 4.4 (2008): 1–15.
Snyder, John. *Prospects of Power: Tragedy, Satire, the Essay, and the Theory of Genre*. Lexington: University Press of Kentucky, 1991.
Soper, Kate. *What Is Nature?* Oxford: Blackwell, 1995.
Spivak, Gayatri. *A Critique of Postcolonial Reason: Toward a History of the Vanishing Present*. Cambridge: Harvard University Press, 1999.
Stavans, Ilan, ed. *Cesar Chavez: An Organizer's Tale*. New York: Penguin, 2008.
Stein, Rachel. "Bad Seed." In *Postcolonial Green: Environmental Politics and World Narratives*, edited by Bonnie Roos and Alex Hunt, 177–93. Charlottesville: University of Virginia Press, 2010.
Streiffer, Robert, and Thomas Hedemann. "The Political Import of Intrinsic Objections to Genetically Engineered Food." *Journal of Agricultural and Environmental Ethics* 18.2 (2005): 191–210.
Swyngedouw, Erik. "Apocalypse Forever? Post-political Populism and the Spectre of Climate Change." *Theory, Culture, and Society* 27.2–3 (2010): 213–32.
Szilard, Leo. "Can We Avert an Arms Race by an Inspection System?" In *One World or None*, edited by Dexter Masters and Katherine Way. London: Latimer House Limited, 1947.
Szilard, Leo. *Voice of the Dolphins and Other Stories*. Expanded ed. Stanford: Stanford University Press, 1992.
Tennyson, Alfred Lord. "Tithonus." In *Norton Anthology of English Literature*, 5th ed., vol. 2, edited by M. H. Abrams, 1110. New York: Norton, 1986.
Teo, Tze-Yin. "Responsibility, Biodegradability." *Oxford Literary Review* 32.1 (2010): 91–108.
Thorpe, Charles. "Violence and the Scientific Vocation." *Theory, Culture & Society* 21.3 (2004): 59–84.
van Wyck, Peter. *The Highway of the Atom*. Montreal: McGill-Queen's University Press, 2010.
van Wyck, Peter. *Signs of Danger: Waste, Trauma and Nuclear Threat*. Minneapolis: University of Minnesota Press, 2005.
Vara-Orta, Francisco. "A Reservoir Goes Undercover." *Los Angeles Times* (June 10, 2008). http://articles.latimes.com/2008/jun/10/local/me-balls10 (accessed February 26, 2015).

Vincent, Bernadette Bensaude. "Plastics, Materials and Dreams of Dematerialization." In *Accumulation: The Material Politics of Plastic*, edited by Jennifer Gabrys, Gay Hawkins, and Mike Michael, 17–29. London: Routledge, 2014.

Viramontes, Helena María. *Under the Feet of Jesus*. New York: Plume, 1995.

Visvanathan, Shiv. "Bhopal: The Imagination of a Disaster." *Alternatives* 11 (1986): 147–65.

Vonnegut, Kurt. *Cat's Cradle*. New York: Dell, 1998.

Vonnegut, Kurt. *Wampeters, Foma, and Granfalloons*. New York: Dell, 1974.

Voosen, Paul. "Geologists Drive Golden Spike toward Anthropocene's Base." *E&E Publishing* (September 17, 2012). http://www.eenews.net/stories/1059970036 (accessed February 26, 2015).

Wallace, Molly. "'A Bizarre Ecology': The Nature of Denatured Nature." *ISLE: Interdisciplinary Studies in Literature and Environment* 7.2 (Summer 2000): 137–53.

Wallace, Molly. "Discomfort Food: Analogy, Biotechnology, and Risk in Ruth Ozeki's *All Over Creation*." *Arizona Quarterly* 67.4 (Winter 2011): 135–61.

Wallace, Molly. "Will the Apocalypse Have Been Now? Literary Criticism in an Age of Global Risk." In *Criticism, Crisis, and Contemporary Narrative: Textual Horizons in an Age of Global Risk*, edited by Paul Crosthwaite, 15–30. New York: Routledge, 2011.

Wallace, Ronald. *The Makings of Happiness*. Pittsburgh: University of Pittsburgh Press, 1991.

Wallace, Ronald. "Poems: 'Toads, and All This Fiddle: Wallace on Music, Metaphor, and Mirth.'" In *Contemporary American Poetry: Behind the Scenes*, edited by Ryan Van Cleave, 292–97. New York: Longman, 2003.

Watts, Jonathan. "Fukushima Parents Dish the Dirt in Protest over Radiation Levels." *Guardian* (May 2, 2011). http://www.guardian.co.uk/world/2011/may/02/parents-revolt-radiation-levels (accessed February 25, 2015).

White, Hayden. "Interpretation in History." *New Literary History* 4.2 (Winter 1973): 281–314.

Whiteside, Kerry. *Precautionary Politics: Principle and Practice in Confronting Environmental Risk*. Cambridge: MIT Press, 2006.

Williams, Terry Tempest. *Refuge: An Unnatural History of Family and Place*. New York: Vintage, 1991.

Willis, Susan. "The Forensics of Spinach." *South Atlantic Quarterly* 107.2 (Spring 2008): 355–72.

Wilson, David. "Despite Boycott, Grape Sales are Up." *New York Times* (November 27, 1988). http://www.nytimes.com/1988/11/27/us/despite-boycott-grape-sales-are-up.html?ref=davidswilson (accessed February 26, 2015).

Yaeger, Patricia. "Editor's Column: The Death of Nature and the Apotheosis of Trash; or, Rubbish Ecology." *PMLA* 123.2 (March 2008): 321–39.

Yaeger, Patricia. "Editor's Column: Sea Trash, Dark Pools, and the Tragedy of the Commons." *PMLA* 125.3 (May 2010): 523–45.

Yamaguchi, Mari. "Fukushima Nuclear Disaster Report: Plant Operators Tokyo Electric and Government Still Stumbling." *Huffington Post* (July 23, 2012). http://www.huffingtonpost.com/2012/07/23/fukushima-dai-ichi-nuclear-plant-operators_n_1694476.html (accessed February 26, 2015).

Yamashita, Karen Tei. *Through the Arc of the Rain Forest*. Minneapolis: Coffee House Press, 1990.
Zaikab, Gwyneth Dickey. "Marine Microbes Digest Plastic." *Nature* (March 18, 2011). http://www.nature.com/news/2011/110328/full/news.2011.191.html (accessed February 26, 2015).
Zins, Daniel. "Rescuing Science from Technocracy: *Cat's Cradle* and the Play of Apocalypse." *Science Fiction Studies* 3.2 (July 1986): 170–81.
Zins, Daniel. "Seventeen Minutes to Midnight." *Nuclear Texts and Contexts* 7 (Fall 1991): 6–10.
Žižek, Slavoj. *First as Tragedy, Then as Farce*. London: Verso, 2009.
Žižek, Slavoj. *In Defense of Lost Causes*. London: Verso, 2008.

Index

Note: Page numbers in italics refer to figures.

9/11, 17–18, 212n63
350.org, 156

activism, 51, 69, 78–82, 84–92, 119–20, 184–85; against GM foods, 100, 110–11; and media staging, 68–70; against plastic grocery bags, 132
Agent Orange, 110–12
Alaimo, Stacy, 128, 132, 148
Alberta Star, 177–79, 181, 183, 184
allegory, 71
Allison, Rachel Hope, *I'm Not a Plastic Bag*, 25, 134, 139–41, 153
American Beauty (1999), 132, 134
anachronism, 3, 51
analogy, 26, 88, 188; and Bhopal, 72, 75, 88–89; for critical practice, 60–61; genome as "book of life," 118; and GM foods, 24–25, 93–97, 100–104, 108–9, 111–22; for information, 114–18; politics of, 68
Andre, Leroy, 179–80
Anthropocene, 127, 142, 156, 188, 192, 206
anthropocentrism, 143
anthropomorphism, 25–26, 129, 148, 229nn29–30; strategic, 129–39, 141–44, 150, 152
apocalypse, 15, 51–52, 89–90, 193–94; fast, 7, 186; imagined scenarios, 57–58; slow, 7, 17–18, 29, 90, 115–17; and world government, 42
archives of risk, 22–23, 28–33, 60–63, 70, 89, 117, 118, 126, 129; and irony, 31–43

atomic pile, 34, 58–62, *61*, 162
atomic scientists, 23, 32–41; "scientists' movement," 41–56, 162, 192. See also *Bulletin of the Atomic Scientists*
Atwood, Margaret, 98

bacteria, 95, 146–47, 150
Bahrani, Ramin: *Plastic Bag*, 25, 134–39, 143
Bainbridge, Janet, 121–22
Bambara, Toni Cade, 54; *The Salt Eaters*, 23, 39–40, 63
Bannet, Eve Tavor, 97, 122
Barthes, Roland, 151–53
Baudrillard, Jean, 212n65; simulacra, 70–71, 73, 74
Beck, Ulrich, 23–24, 131, 154, 164, 187–89; on "becoming-real" of risk, 14, 171; on "boomerang effect," 67; on Chernobyl, 160; on cosmopolitanism, 83, 87, 157; on dialectic of irony, 41, 78; on "enlightenment" function, 53, 56; on experts, 95; and first modernity, 128; on insurance, 116; on poisoning, 77; on political activism, 85–87, 111, 138; reflexive second modernity, 18, 30–33, 56, 197; *Risk Society*, 10, 15; risk society model, 2, 4, 10–15, 17, 27, 33, 37, 56, 62, 71, 120–21, 174; on satire, 35, 232n10; on staging of risk, 66, 68–69, 99–100; on subpolitics, 41, 56, 69, 78, 83, 120, 138; on truth, 19; on uncertainty, 19, 125; *World at Risk*, 10, 212n63

Belgian Congo, 233n31
Bell, Daniel, 43
Bennett, Jane, 129, 131, 133, 134
Bérubé, Michael, 75
"Bhopal gesture," in *White Noise*, 24, 72, 74–75, 82, 89, 92
Bhopal toxic chemical disaster, 2, 24; and analogy, 94; death toll, 73, 218n31; as figure, 87–89; as metaphor, 65–70; ongoing nature of, 82; treatment of victims, 73, 85, 219n36; "we all live in Bhopal" slogan, 65–68, 73–74, 89–90, 92, 222n116
Bikini atoll, nuclear testing on, 2, 54, 64–69, 71–72, 90, 92, 94
Bill and Melinda Gates Foundation, 116
biodegradability, 123, 126–27, 136, 147
biopower, 92
biotechnology, 2, 10, 21, 24–25, 192; corporate control over, 100, 105–9, 114–17, 120–21 (*see also* Monsanto); patent protection, 103–4. *See also* genetically modified foods
Blondin, George, 166, 176
Blow, Peter, *Village of Widows*, 26, 162, 164–69, 175, 184
Body 2.0: Creating a World That Can Feed Itself ("Zeitgeist" forum), 113–17
Borlaug, Norm, 113
boycott politics, 69, 85, 111, 120; UFW grape boycott, 79–81, 220n59
Bradford, George (pseudonym), 72, 84, 91, 92; "We all live in Bhopal," 24, 65
Bradley, David, 57–58, 64–65, 68, 91–92
Brand, Stewart, 26, 155
Brazil, 106–7, 141–46
Brians, Paul, 8
Brilliant, Larry, 113–15
Brown, Bill, 138
Bt plants, 110, 112, 224n26
Buell, Frederick, 15, 17, 71, 75

Buell, Lawrence, 8–9, 15, 64, 65, 83, 210n32; "toxic discourse," 69, 71, 99–101
Bulletin of the Atomic Scientists: founding of, 43, 45; "second nuclear age," 2–3, 9, 12, 25–26, 29, 55, 153, 156, 192–94, 206. *See also* Doomsday Clock
Burke, Kenneth, 23, 31, 214n44
Bush, George W., 19, 28, 49, 95

Cameco corporation, 159, 161, 176, 181
Canada, 224n30; Global National TV, 175; government responsibility for uranium contamination, 165–67, 175–76; tar sands oil extraction, 184–86; uranium mining, 26, 29, 159, 162–83, 187, 189, 191
Canada-Déline Uranium Table Report, 177, 179–80
cancer, 154, 223n9; compensation for victims, 189; and radiation, 164, 171; rates near Port Radium, 163, 166, 177; risk of, 123–24
Carruth, Allison, 114
Carson, Rachel, 37, 222n118; *Silent Spring*, 8, 46, 71, 83–84, 99, 224n26
catastrophes, 44, 90–91; emergence of, 33; location of, 65–66, 74–76; potential, 94; slow, 163; staging the art of, 197–204
cause and effect, 154–55, 163–65, 171
Chakrabarty, Dipesh, 17
Chaloupka, William, 6
chance, 66–67, 74
Chávez, César, 79–81, 92, 220n59
Chernobyl nuclear accident, 14, 78–79, 88–89, 177, 217n14; as analogy, 94, 122; compared to Fukushima, 158–60; quarantine of, 35
China, 93, 114, 176
Clark, Timothy, 123
Clements, Marie, *Burning Vision*, 26, 162, 169–75, 180, 184–87, 236n100
climate change, 2, 10, 16; and agriculture, 113–14, 116–17; as cinematic,

191; and consumers, 138; as current catastrophe, 66; disproportionate effects of, 186; reality of, 20; relationship to the nuclear, 25–26, 206; and responsibility, 190; as synecdoche, 211n40
Climate Corporation, 116–17
climate denial, 160
Cohen, Judith Beth, 101, 109, 226n69
Cohen, Tom, 158, 161, 191, 211n40
Cold War: beginning of, 43; end of, 1, 7, 10, 28–29, 206; late period, 195; and nuclear criticism, 4; nuclear fears, 17, 22, 36–37, 42–43, 66, 158
colonialism, 144–45
consumerism, 78–81, 120, 144
consumers, 136–38; environmentalist, 132, 140
contamination: fear of, 125; responsibility for, 39–40
Cooper, Ken, 38–40, 54, 57, 60
Cordle, Daniel, 6, 9
Corwin, Jeff, 140–41
cosmopolitanism, 11, 83
Council for a Livable World, 46
Coupland, Tim, 177–80
critical practice, metaphors for, 60–62
cross-contamination, 204, 206
cross-pollination, 101, 105, 121
Crutzen, Paul, 192
Cuban missile crisis (1962), 35, 46
Cummings, Claire Hope, 95

data, genetically modified seeds as, 114–17, 122
deconstruction, 3, 5, 19–20
DeLillo, Don, 28–29; *Underworld*, 28, 57, 75, 152; *White Noise*, 15, 24, 69–83, 89, 92, 94, 120, 127, 147
Déline, Canada, 163–69, 178, 187; "The Water Heart," 186
Déline Uranium Committee, 165–67, 176, 178, 185
DeLoughrey, Elizabeth, 57
de Man, Paul, 66, 126, 129, 135, 147
Dene people, 163–69; prophesies, 171–74, 180, 186–87

depleted uranium, 10, 29, 51, 57
Deresiewicz, William, 35
Derrida, Jacques, 3–5, 13, 18, 28, 33, 187, 188; on archives, 33, 45; on biodegradability, 123, 126–27, 136, 147; on consigning, 30; on futurity, 190; on nuclear war, 3–8, 10, 17; on post-9/11 world, 17–18; on reflex and reflection, 26–27; on responsibility, 192; on textuality, 15
deterrence, 6, 7, 43, 71; rhetoric of, 4
Dickinson, Adam, "Hail," 25, 134, 149–51
disaster capitalism, 20
discourse analysis, 3, 5, 19–20
documentary films, 82, 107, 163, 184
Doomsday Clock, 1–4, 6, 8, 16–18, 21–22, 27, 45, 55, 156, 192–94
doomsday scenarios, 193–94; and seed vaults, 115–16, 122
Dow Chemical (formerly Union Carbide), 76–77, 82, 85, 87, 110
Dower, John W., 197
DuPont, 59
Duvall, John, 78
Dyson, Freeman, 93, 94

earthquakes, 182–83
ecocriticism, 4, 71, 83, 210n32; emergence of, 8–10; privileging of the local, 67–68
Ehrlich, Paul, 91, 113, 222n113, 222n116
Einstein, Albert, 44, 56
ekphrastic poetry, 26, 198, 200–204
Eldorado Limited, 159, 176, 181
environmental catastrophes, "slow violence" of, 62–63, 69
environmental justice, 10, 54, 68, 163; and racial politics, 39
ethics, 36, 59, 140
Eubanks, Charlotte, 195, 200–201
Ewald, François, 21, 173
extrapolative fiction, 98

"fabulously textual," 5, 9–10, 13–14, 20, 23, 210n22

258 INDEX

farmworkers, 79–82
fear, 56, 86; consumer, 81; of genetic engineering, 96
Ferguson, Frances, 14
Fermi, Enrico, 23–24; atomic pile, 58–62; in Millet's work, 44–45, 53; wager on nuclear risk, 34–36, 40, 57
Fieser, Louis, 36
first nuclear age, 1–4, 29; and archive of global risk, 32–47; experts, 10; origins of, 26, 33–35, 60–63; radiation risk, 158; scientists' movement, 41–56; as template for incalculable risk, 41
food scarcity, 113, 223n9
Fortun, Kim, 1, 76, 219n45; *Advocacy after Bhopal*, 76
Foucault, Michel, 23, 30–31, 38, 46, 92
Fowler, Cary, 115
fracking, 192
Franck report, 43
"Frankenfoods," 96, 98, 105
Fukushima nuclear accident, 2, 29, 35, 181–83, 189; cleanup, 182–83; contaminated areas, 156, 182–83, 187–88; in context of first nuclear age, 158–60; regulations on radiation exposure, 187–88; responses to, 11
future anterior, 22, 147, 171

Garbage Patch. *See* Pacific Garbage Patch
Garrard, Greg, 121
Gates Foundation, 116
genetically modified (GM) foods, 24–25; discomfort with, 24, 93–97, 121–22; extrinsic economic objections, 96, 105–9, 114; intrinsic moral objections, 96, 101–5, 108, 114, 121; uncertain risks of, 97–101, 109–12, 223n9
genetically modified organisms (GMOs), 3, 118
genetic use restriction technology (GURT), 103, 107
Genuity™ soybeans, 93

global risk, 22, 23–24, 192–94, 206–7; archive of, 30, 32–33; cumulative nature of, 59–60, 62, 65–66; incalculable, 34; nuclear as model for, 29; origins of, 33–35, 57, 62–63; uneven distribution of, 12, 39, 66–68, 186. *See also* risk
global warming, 19–20. *See also* climate change
GM foods. *See* genetically modified foods
GMOs. *See* genetically modified organisms
Godmilow, Jill, 167–69
Golden Rice, 112–13
Google, 25, 113, 115–16
Gore, Al, 138
Grandey, Jerry, 176, 181–82
Grant, Hugh (Monsanto CEO), 25, 113–16, 119
grape boycott, 79–81, 220n59
Grausam, Daniel, 215n87
Great Bear Lake, Canada, 162, 169; fishery management, 180; preservation of, 186; Working Group, 180
greenhouse gases, 155, 157, 232n7
Greenpeace, 69, 86, 177; Shell protest, 138
green revolution, 65, 91, 96, 113
grocery bags, plastic, 132–40, 153
grocery stores, 75, 79–81
Guha, Ramachandra, 84

Hamblin, Jacob Darwin, 160–61, 177
Haraway, Donna, 102, 104, 106
Hardt, Michael, 105–6, 108–9, 121
Hawking, Stephen, 55–56
Hawkins, Gay, 25, 132, 134, 141, 148
hazards, 2–3, 10–12; environmental, 17; unseen, 66, 183. *See also specific hazards*
HDPE (high-density polyethylene), 125–26, 152
Heal the Bay, 132–36
Heffernan, James, 201–2
Heise, Ursula, 11, 15, 67, 75, 77, 81–82, 142–43, 217n3, 226n69

Hendrix, John, 193–94
Henningson, David, *Somba-Ke*, 26, 162, 164, 175–84, 187
herbicides, 37, 84, 91, 110
Herring, Ronald J., 106–8, 110, 226n62
Herzog, Werner, 135–36
Hird, Myra, 142
Hiroshima: aftermath of atomic bombs, 26; artwork on (*see* Maruki, Iri and Toshi; Wallace, Ronald); bombing of, 7, 35, 65, 195–97; ground zero, 51; lantern ceremony, 167, *168*; visit by Déline Uranium Committee, 165–67, 175, 185
Hyde Park, Chicago, 58–60

immigration, 108
immortality, 137–38
India, genetically modified seeds in, 107
inequality, 192; in distribution of global risk, 12, 39, 66–68, 186
INES (International Nuclear Event Scale), 158–60
information: genetically modified seeds as, 114–18, 122; multiplication of, 61–62
Institute, West Virginia, 75–77, 82, 219n45
insurance, 116–17
International Atomic Energy Agency (IAEA), 158, 160–61
International Campaign for Justice in Bhopal, 68
internationalism, 22, 55
Iovino, Serenella, 130
Iran, 2, 28
Iraq war, 20, 29, 49, 119
irony, 26, 52; and dialectic, 214n44; dialectics of, 41–43; generalized, 32; and politics, 48–49; of risk, 23, 31–43, 60; in *White Noise*, 77, 79, 83
Ivanhoe Reservoir, Los Angeles, 123–26, 140, 152–53

Jakobson, Roman, 90
Japan: atomic bombing of, 7, 26, 35, 43, 48, 142, 159, 170–71; nuclear industry, 161, 176. *See also* Fukushima nuclear accident; Hiroshima
Jasanoff, Sheila, 33–34, 36, 40, 41, 43–44, 57–58, 76, 82, 92
Johnson, Barbara, 229n29
Jordan, Chris, 128, 150
Junkerman, John, 203

Kasperson, Jeanne X., 11
Keen, Sam, 49–50
Kermode, Frank, 52
Keystone XL pipeline, 159
Kimbrell, Andrew, 93, 94, 96, 97
Klein, Naomi, 20, 156–57
Klein, Richard, 6, 13–14

LaBine, Gilbert, 166, 185, 234n47
Ladino, Jennifer, 218n24, 225n55
Lamb, Jonathan, 135–37
Lamorisse, Albert, 230n43
Latour, Bruno, 3, 5, 18–21, 60–62, 131, 146
Le Guin, Ursula K., 98
Leiss, William, 11
lettuce boycott, 79
Liboiron, Max, 128
listing, rhetorical technique of, 130–31
literary studies: comments on genetic technologies, 95; subfields of, 4; turn to objects and things, 130–31
Los Alamos, 38, 46
Los Angeles Department of Water and Power, 123–24, 153
Los Angeles reservoir, plastic balls in, 123–26, 140, 152–53
Lovelock, James, 26, 155
Luckhurst, Roger, 10
Lucky Dragon (ship), 196, 216n99

The Majestic Plastic Bag (Heal the Bay), 132–36
Manhattan Project, 44, 46, 176, 233n31
Mansfield, Nick, 187, 188
Martucci, Elise, 218n32

Maruki, Iri and Toshi, 26, 194–207, 203; *Atomic Desert*, 198, *199*; *Hell*, 196–97, 203–4
Maruki, Toshi, *Hiroshima No Pika*, 195–96, 207
matter, liveliness of, 131, 133–34, 153
May, Theresa, 169, 174, 175
McHugh, Susan, 119, 223n11
McKibben, Bill, 155, 156
McMurray, Andrew, 1, 17, 212n65
megarisks, 10, 12, 17, 20, 28–29, 66
messianism, 42–43, 45; end-times, 49–50, 52; scientific, 54–55
Messmer, Michael, 71–72, 78–79
Metallurgical Lab (Chicago), 46, 59, 61
metaphor, 23–24, 26, 90; airborne toxic event as, 71; Bikini as, 64; politics of, 68, 121
metonymy, 23–24, 26, 46, 65, 90; Bhopal, 68, 73; Bikini, 64; distinction between metaphor and, 66; politics of, 68
MIC (methyl-isocyanate) gas, 65, 67, 68, 75–76, 82, 91, 219n36
Middleport, New York, 76, 82
Millet, Lydia, *Oh Pure and Radiant Heart*, 17, 23, 44–55, 192
Minamata, Japan, 26, 88, 196
miscegenation, 103, 108
Mitchell, W. J. T., 204
Mizruchi, Susan, 91
modernity, 91–92; first, 128, 155; second, 11, 18, 30–33, 56, 155, 197
Monsanto corporation, 24–25, 98, 103, 107–8, 110–17, 223n9, 224n26, 224n30
monsters, 102, 105–6, 108, 134, 137, 141
Moore, Patrick, 26, 28–29, 155–56, 177, 232n10
moral objections to GM foods, 96, 98, 101–6, 108, 121
Morton, Timothy, 93, 94, 96–97, 105–6, 108, 127, 130, 148, 154–55
Muecke, D. C., 32, 34–35
mutations, 145

mutually assured destruction (MAD), 4, 30, 43, 195

Nagasaki, 7, 26, 35
National Oceanic and Atmospheric Administration (NOAA), 127–28, 151
natural resources, 143–46. *See also* plastic; uranium mining
nature: and genetic modification, 101–6; nature/culture binary, 131–34
Navajo (Dine) country, 178
Nazi Germany, 11, 44, 56
Negri, Antonio, 105–6, 108–9, 121
neocolonialism, 144–45
Nesbitt, Lorien, 186
NewLeaf™ potatoes, 98, 110, 111–12, 114, 224n26
new materialisms, 129–30
nihilism, 52, 135, 232n10
Nikiforuk, Andrew, 179
Nixon, Rob, 22, 62, 64, 83–84, 88–89, 92, 143, 164
NOAA. *See* National Oceanic and Atmospheric Administration
nonbiodegradability, 126; of plastic, 137–42, 147
No Place to Run (1985), 68, 75, 82
Norris, Christopher, 9–10
Norris, Margot, 46
North Korea, 2, 9, 28, 43
nostalgia, 52, 83
nuclear criticism, 3–4; as anachronistic, 10; and the end of time, 4–7; in second nuclear age, 8–9
nuclear denial, 160, 164, 178
Nuclear Energy Institute, 156, 158
nuclear fallout, 91; Cold War fears of, 222n118
nuclear industry, 155–56, 176
nuclear power: as "clean and green," 155–58, 181–82, 232n7; supporters, 155–58, 176–77
nuclear risk, 1–4, 28–33; accidents and insurance, 14; futurity of, 66; global reach of, 64; as psychologi-

cal, 177–78. *See also* first nuclear age; second nuclear age
nuclear war: textuality of, 5–7; total thermonuclear war, 4–6, 8, 17, 38, 42, 206
nuclear waste, 28–29; durability of, 126–27; storage, 188–89; uses, 57
nuclear weapons, 28–29; allegory about, 35–37; "Olympian" story of, 38, 40, 60; as "peril and hope," 42–44, 46, 55, 62

Obama, Barack, 16, 28
object narrative, 25. *See also* things
O'Brien, Susie, 181
OECD (Organization for Economic Co-operation and Development), 97
oil industries, 184–86
One World or None (1946), 23, 42
Oppenheimer, J. Robert, 23–24, 28, 41–44, 46, 55–56, 62; in Millet's work, 43–46, 49, 50, 52–53; at Trinity explosion, 33, 60
Oppermann, Serpil, 130–31
Osteen, Mark, 72–74
Ozeki, Ruth, 24–25; *All Over Creation*, 24–25, 98–105, 107–14, 116–22; *My Year of Meats*, 99, 101

Pacific Garbage Patch, 2, 25, 125–28, 138–42
Packer, Matthew, 70
Palmer, James, 135, 137
peace, and nuclear weapons, 42–43
"peril and hope," 42–44, 46, 55, 62
Perrow, Charles, 11, 160, 178
personification, 129–30, 229nn29–30
pesticides, 37, 84, 91–92, 220n59; and biotechnology, 224n26; in Bt plants, 110; dangers of, 79–81, 94, 99, 118, 222n118; pervasiveness of, 68, 80–81
PET plastic, 125
Petryna, Adriana, 88
pharmakon, 3, 42
Phillips, Dana, 121

plastic, 2, 25, 192; grocery bags, 132–40, 153; materiality of, 130–34; morphability of, 151–53; pervasiveness of, 148–51, 153; plasti-morphism, 148; representations of, 127–28; reservoir balls, 123–26, 140, 152–53; toxicity of, 123–30
plastic waste, 127, 129, 136, 139–53; classified as solid waste, 152; non-biodegradability of, 137–42, 147; recycling, 152; wildlife consumption of, 128, 139, 148–51. *See also* Pacific Garbage Patch
political activism. *See* activism
Pollan, Michael, 25, 102, 113–14
pollutants, 29, 31, 125, 128, 132, 145, 152, 158, 164, 183. *See also* Pacific Garbage Patch; plastic waste; radiation; toxic chemicals
population growth, 91, 113
Port Hope, Ontario, 162–63
Port Radium, Canada: as paradigm, 183–86, 190; uranium mining, 26, 162–69, 176–81
postmodernism, 70, 78
poststructuralism, 4, 8–9
poverty: and GM seeds, 107; and risk, 73, 77–79, 81, 84
precaution, 46, 117, 119, 147–48, 163, 188–90
precautionary reading, 18–27, 94, 122, 206
prosopopoeia, 129–30, 134, 139, 229nn29–30
purity: of plant genes, 103, 105–7; racial, 102–3
Pusztai, Arpad, 223n9

Rabinowitch, Eugene, 43, 48, 56, 58, 162
radiation: denial of risks, 160–61, 164; durability of, 126–27; exposure, 13, 57–60, 64–66, 145, 162, 217n14; invisibility of, 183; poisoning, 216n99; risks, 158–60, 177–78
Radiation Exposure Compensation Act (U.S.), 189–90

INDEX

Rapture, 45, 50, 52
the real: ecocriticism's privileging of, 8–9; referents, 15; and risk criticism, 18–19
Ream, Walt, 95–96
Rees, Martin, 2
referent, 5–7, 14–15
reflexivity, 23, 26–27, 44, 46, 156, 194–95; in second modernity, 18, 30–33, 56, 197
regulation: chemical plant safety regulations, 67, 76, 92; of GM foods, 96–97, 100–101, 114, 119, 122, 224n30; of plastics, 152; on radiation exposure levels, 187–88
religious views of environment, 101–5, 225n55
representation: of plastic, 127–28; of time, 6
resilience, 20, 137, 147
resource extraction, 143–46, 180. *See also* uranium mining
responsibility, 39–40, 188–90, 192, 204; abdication of, 161, 188; Canadian government, 165–67, 175–76; difficulty of identifying, 164; model for taking, 185; and uncertainty, 173
Rhodes, Richard, 34, 36
risk: "becoming-real," 14, 18, 24, 62, 66, 94, 171, 173, 194; distinction between catastrophe and, 66; incalculable, 20, 33–38, 41, 57–58, 63, 68, 109, 188–90; "real referent," 13; "stigmatizing" of, 11; textuality of, 13–15. *See also* global risk
risk assessment, scientific, 19–20; limits of, 13
risk/benefit analysis, 14; GM agriculture, 109–11; nuclear power, 158, 181
risk communication, 11, 16–17
risk criticism, 3–4, 10–15, 26–27, 206–7; atomic pile as metaphor for, 60–63; paradigmatic examples of, 163; as reflexive, 156, 194–95
risk management, 183; discourse, 10–12, 31–32, 187–88; simulacral logic of, 70
risk society, 186–91; Beck's model of, 2, 4, 10–15, 17, 27, 33, 37, 56, 62, 71, 120–21, 174; "becoming real" of, 173; satire of, 75
Roberts, Jody, 127–28
Robin, Marie-Monique, 107, 115
Rody, Caroline, 142
Rollin, Bernard, 96
Roundup Ready® plants, 93, 110, 112
Rumsfeld, Donald, 20
Ruthven, Kenneth, 6, 8, 10

Sahtu Dene people. *See* Dene people
Sandilands, Catriona, 184–85
satire, 44, 52
Schell, Jonathan, 40, 65
Schmeiser, Percy, 224n30
Schwenger, Peter, 5–6, 8, 210n22
science: applied, 118; and doctrine of substantial equivalence, 97; indictment of, 48–49
science fiction, 47, 98
scientific certainty, lack of, 19–21, 173
scientists, 30–31, 213n22; and morality, 36, 38, 41, 46, 59, 214n42. *See also* atomic scientists; scientists' movement
scientists' movement, 41–56, 162, 192
sea life, plastic consumption by, 128, 139, 148–51
second nuclear age, 2–3, 8–10, 12, 16–18, 23, 25–26, 29, 46, 55–57, 153, 156, 158, 192–94, 206
seeds, genetically modified: corporate control over, 103–8; as data or information, 114–18; piracy, 106–8
SENES, 177
Seren, Leo, 59, 65
Shell, 138; boycott, 69
Shiva, Vandana, 94, 104
Sibree, Bron, 87
Silent Spring (Carson), 8, 46, 71, 83–84, 99, 224n26
simile, 97, 134, 155

Sinha, Indra, *Animal's People*, 24, 69–70, 82–91, 94, 120
Slovic, Paul, 11
slow violence, 83, 88, 164, 171, 183, 186–87
Smith, Annie, 184–85
Smith, Jeffrey M., 223n9
Snow, C. P., 62
Snyder, John, 51
Soper, Kate, 9
Soviet Union, 1, 43, 48, 88
speculative fiction, 98
staging of catastrophe, 73–75, 77–79, 83–87, 195, 197–204
staging of risk, 99–100, 163, 194; in environmental literature, 184–85; in films, 191; GMOs, 119–21; by industry, 188; in media, 68–70
StarLink™ corn, 93
Stavans, Ilan, 79
Streiffer, Robert, 101
strontium 90, pervasiveness of, 35, 67, 217n14
subpolitics, 41, 56, 69, 78, 83, 117, 120, 138, 167
substantial equivalence (of GM foods), 97, 100–101, 111, 114, 121–22
supermarkets. *See* grocery stores
Svalbard seed vault, 25, 115–17, 122
Swyngedouw, Erik, 157
Sygenta, 116
synecdoche: Bhopal, 73, 88; Bikini, 64; Doomsday Clock, 2; and GM foods, 97, 114; second nuclear age, 23, 29
Szilard, Leo, 23–24, 56; "Has the Time Come to Abrogate Nuclear War," 47; in Millet's work, 44, 46–49, 53; "My Trial as a War Criminal," 48; "Report on 'Grand Central Terminal,'" 49, 54; "The Mined Cities," 47, 49; *The Voice of the Dolphins*, 23, 47–49, 54

tar sands oil extraction, 184–86
Teller, Edward, 145
Temik, 65, 217n8, 219n45
temporality. *See* time
Teo, Tze-Yin, 126
TEPCO (Tokyo Electric Power Company), 159, 161, 182, 235n91, 236n95
Terminator technology, 103–5, 107, 111
textuality, 5, 9–10, 13–14, 20, 23, 210n22
theological objections to GMOs, 101–5
things, 130, 135–37, 139–40, 143, 149
thinking machines (computers), and metaphor of atomic pile, 60–61
Thorpe, Charles, 42–43
Three Mile Island nuclear accident, 72
time: end of, 4–7; messianic end-times, 49–50, 52; nuclear, 7; of representation, 6; temporality of plastics, 123–30
total thermonuclear war, 4–6, 8, 17, 38, 42, 206
toxic chemicals, 2, 70–92; accidental spills, 2, 24, 65–66; agricultural, 24; and conventional agriculture, 109–12; invisibility of, 183
toxic discourse, 69, 71, 99–101, 122
Trinity nuclear test explosion, 40–41, 44, 50, 58, 72, 162; fetishizing "ground zero" of, 38, 54; as origin of nuclear age, 33–34, 38, 56–57, 60; as public demonstration, 43
tropes, 26; master tropes, 23, 31
tropological wagers, 26, 31, 66, 88, 118, 129–30, 206; analogy, 88, 93–97; description of, 23
tsunami, 183, 236n95
Turing, Alan, 60–61

uncertainty, 20–27; lack of scientific certainty, 19–21, 173; of precautionary principle, 173; risks of GM foods, 96–101, 109–12, 114, 119–22

Union Carbide: Bhopal, India plant, 2, 24, 65, 67, 73–77, 82, 90–92; boycotts, 85; and nuclear industry, 91; safety regulations, 67, 76, 92; West Virginia, US plant, 75–77, 92. *see also* Dow Chemical

United Farm Workers (UFW) boycott, 79–81

United Nations, 42, 50

United States: imperialism, 38; nuclear reactors, 236n96; Nuclear Regulatory Commission, 182; nuclear testing, 1; nuclear weapons, 2; pesticide risks, 24, 76; Radiation Exposure Compensation Act, 189–90; risk in, 74; toxic chemicals in, 74–82

uranium mining, 2–3, 233n31; in Canada, 26, 29, 159, 162–83, 187, 189, 191; depleted uranium, 10, 29, 51, 57; and government responsibility for cleanup, 165–67, 175–76

van Wyck, Peter, 13–14, 17, 26, 127, 163, 165, 169, 186, 187, 189–91, 233n38, 234n47

Vietnam, 43, 110–11

Viramontes, Helena María, 80–81

Visvanathan, Shiv, 92

Vonnegut, Bernard, 213n22

Vonnegut, Kurt, 23, 35–41, 53, 214n42, 232n10

wagers, 26, 32, 35–43, 57–58, 60; of irony, 41–43; satirical critique of, 35–39. *See also* tropological wagers

Wallace, Peg, 238n21

Wallace, Ronald, "The Hell Mural," 200–207

waste disposal, 142. *see also* plastic waste

Waste Isolation Pilot Plant (WIPP), 189

Watson, David. *See* Bradford, George

White, Hayden, 23, 31

whiteness, 38–39, 54, 70

Whiteside, Kerry, 21, 93, 94, 96, 97, 122

wildlife, plastic consumption by, 128, 139, 148–51

Williams, Terry Tempest, *Refuge*, 99, 154–55, 189, 190–91

world government, 50–51

Yaeger, Patricia, 131, 133, 146

Yamashita, Karen Tei, *Through the Arc of the Rain Forest*, 25, 134, 141–48, 151

Yucca Mountain, 29

"Zeitgeist" forum, 113–17

Zins, Daniel, 8, 35–36

Žižek, Slavoj, 20, 22, 108, 161